Distributed Energy Resources and Electric Vehicle

Explore the prospective developments in energy systems and transportation through an in-depth examination of *Distributed Energy Resources and Electric Vehicle: Analysis and Optimisation of Network Operations*. This innovative publication explores the realm of renewable energy, electric vehicles, and their influence on network operations, offering valuable perspectives for readers from diverse disciplines.

This extensive publication delves into the complex interplay between distributed energy resources (DERs) and electric vehicles (EVs), as well as their incorporation into established power grids. The subject matter encompasses a diverse array of topics, encompassing the attributes and advantages of distributed energy resources (DERs) and electric vehicles (EVs), obstacles related to grid integration, efficient allocation of resources, and strategies pertaining to demand response. The book offers a comprehensive exploration of system analysis and optimisation techniques, emphasising the effective utilisation of distributed energy resources (DERs) and electric vehicles (EVs) in energy networks. It aims to equip readers with a robust comprehension of strategies to optimise the performance and potential of DERs and EVs in this context.

The book focuses on pioneering research and innovative solutions that are at the forefront of enhancing network operations. The authors demonstrate the novelty and applicability of their findings through the examination of real-world case studies and the utilisation of sophisticated mathematical models. This book serves as a highly valuable resource for individuals engaged in research, engineering, policy-making, and industry-related activities who are interested in effectively navigating the dynamic realm of energy systems and transportation. It equips them with the necessary knowledge and insights to make well-informed decisions that contribute to the attainment of a sustainable future.

Distributed Energy Resources and Electric Vehicle

Analysis and Optimisation of Network Operations

Edited by
Aijaz Ahmad, Kushal Jagtap, and
Keerti Rawal

CRC Press
Taylor & Francis Group
Boca Raton London New York

CRC Press is an imprint of the
Taylor & Francis Group, an **informa** business

First edition published 2024
by CRC Press
6000 Broken Sound Parkway NW, Suite 300, Boca Raton, FL 33487–2742

and by CRC Press
4 Park Square, Milton Park, Abingdon, Oxon, OX14 4RN

CRC Press is an imprint of Taylor & Francis Group, LLC

ISBN: 978-1-032-31872-1 (hbk)
ISBN: 978-1-032-31873-8 (pbk)
ISBN: 978-1-003-31182-9 (ebk)

DOI: 10.1201/9781003311829

Typeset in Times
by Apex CoVantage, LLC

Contents

Chapter 5

Anamika Das, Ananyo Bhattacharya, and Pradip Kumar Sadhu

Chapter 10 Analysis of an IUPQC Device Using Conventional PID
and FOPID Controllers in a Wind Energy Conversion System........ 202

*G. Pandu Ranga Reddy, Kushal Jagtap, Y. Chintu Sagar,
and R. Sheba Rani*

Chapter 11 Photovoltaic-Based Battery-Integrated E-Rickshaw with
Regenerative Braking Using Real-Time Implementation 223

Arpita Basu and Madhu Singh

Foreword

Many countries around the world are motivated to work towards resilient and sustainable infrastructure to shape an ecosystem that connects the real world with the digital world. Moreover, the population growth and subsequent health emergency due to air pollution has forced countries to provide reasonable services and switch over to smart city options. Consumers need to be empowered to make their energy systems and processing in buildings and industries more efficient and sustainable. The electrical transportation system is going to be a major energy consumption component in smart cities, as electric vehicles are the key technology to decarbonize road transport, a sector that accounts for 16% of global emissions. Accordingly, the power dispatching operation mode is expected to undergo great changes. There has to be complete coordination of the transmission and distribution network and the microgrid. The charging of electric vehicles has to be managed in a methodical manner so that there will be minimal negative impacts on the distribution network. Consequently, there has to be an effective scheduling control strategy for electric vehicle charging and discharging. A strategy needs to be developed so that a regional electricity load plan is established to stabilize the concerned distribution network. For this, the randomness of electric vehicle charging load needs to be reflected accurately in the operation characteristics of the distribution network. Distributed energy sources in the form of solar photovoltaic and wind energy have a great potential to be exploited for supporting the distribution network in general and the charging circuit in particular.

Reducing petroleum usage for transportation could be a key pathway to reducing emissions. The net effect of EV charging on emissions, both from the power grid and from fuel combustion in an internal combustion engine, varies by system or region based on several factors that primarily include the generation mix and the time of day that vehicles recharge.

This book brings together complicated investigative and applied material on the subject. It will be of great value to engineers, post-graduate students, and industrialists who wish to learn about the details in the area of analysis and optimization of network operations for distribution networks containing charging stations at different locations.

Aijaz Ahmad
Srinagar, India

Preface

Across the globe, there is a notable shift occurring towards sustainable and decentralized energy generation, leading to an increased prevalence of distributed energy resources (DERs) such as solar photovoltaic energy, wind turbine systems, and energy storage devices. Concurrently, the rise of electric vehicles (EVs) as an environmentally friendly mode of transportation has brought about a new era of energy management, filled with challenges and opportunities.

The integration of DERs and EVs holds the potential to reshape the generation, consumption, and management of electricity. However, this integration also presents intricate technological, operational, and optimization-related hurdles that must be overcome to establish a streamlined and efficient energy ecosystem.

This book aims to provide a comprehensive examination of the interactions between DERs, EVs, and network operations, with a specific focus on analysis and optimization aspects. Different contributors explore various methodologies, techniques, and algorithms that enable the effective integration and coordination of these resources, ensuring grid stability, reliability, and economic viability.

Throughout the book, contributor guide readers through a step-by-step journey to grasp the underlying principles, emerging trends, and advanced approaches utilized in the field. Starting with a high-level overview of DERs, EVs, and their impact on the power system, we delve deeper into the complexities of network analysis, modeling, and optimization.

Key topics covered include demand response, grid integration of renewable energy sources, wireless EV charging technologies, optimal charging and discharging strategies, and intelligent energy management systems. The contributors provide a depiction of the pragmatic outcomes and advantages of incorporating DERs and EVs into network operations by means of empirical instances, as well as the utilization of modeling and simulations of practical models.

Furthermore, contributors address the challenges associated with this integration process, including grid stability, power quality, regulatory frameworks, and market mechanisms. By providing insights into these obstacles, the editors' aim is to empower researchers, engineers, policymakers, and industry professionals to make informed decisions and develop innovative solutions in this rapidly evolving landscape.

The editors hope that this book serves as a valuable resource, providing you with a comprehensive understanding of the analysis and optimization of network operations when integrating distributed energy resources and electric vehicles. By embracing these disruptive technologies and applying the knowledge shared within these pages, the editors aim to pave the way towards a sustainable, resilient, and intelligent energy future.

Editors

Aijaz Ahmad received his B.E. (electrical engineering) degree from the National Institute of Technology (NIT), Srinagar, India, in 1984 and his M.Tech. and Ph.D. from the Indian Institute of Technology, New Delhi, India, in 1991 and 1998, respectively. He was lecturer and later assistant professor in the Electrical Engineering Department, NIT Srinagar. Since 2006, he has been working there as a professor. In between, he remained Head of the Department from 2012–2015. His main research interests are power system operation and optimization, power system restructuring and deregulation, flexible AC transmission, and energy system planning and auditing. Aijaz Ahmad is member of the IEEE, fellow of the Institution of Engineers (India), life member of the Indian Society for Technical Education, and Member of the Global Science and Technology Forum.

Kushal Jagtap received his B.E. (electrical electronics and power Engineering) degree from the Government College of Engineering Aurangabad, Maharashtra, India, in 2007 and his M.Tech. (power systems) and Ph.D. from the Indian Institute of Technology Roorkee, India, in 2012 and 2018, respectively. His Ph.D. thesis title was "Loss Allocation in a Radial Distribution System with Distributed Generation." He joined as a postdoc fellow under the project title "Impact of Electric Vehicles on Demand Side Management" at the Indian Institute of Technology Madras, India, in 2018. He has been working as assistant professor in the Electrical Engineering Department at NIT Srinagar since 2018. His main research interests and major topics taught during the last three years are loss allocation, distributed generation, power system operation and optimization, power system restructuring and deregulation, and flexible AC transmission. Kushal Jagtap has been a member of the IEEE for the last three years.

Keerti Rawal received her B.Tech (electrical engineering) from National Institute of Technology Jaipur, Rajasthan, India, in 2013 and her M.E. (electrical engineering) from the Indian Institute of Science Bangalore in 2015. Keerti Rawal is currently pursuing a Ph.D. in the Department of Electrical Engineering at NIT Srinagar. She is exploring the impacts of the integration of renewable energy resources and storage on the electrical grid, primarily focusing on electricity markets and energy analytics in her doctoral research. Her key interests are energy markets, power system optimization, energy analytics, distributed optimization, energy forecasting, electric vehicles, control strategies, spinning reserves, and energy storage technologies. She has been a student member of the IEEE for four years.

Contributors

Aijaz Ahmad
Department of Electrical Engineering
National Institute of Technology
Srinagar, Hazratbal Srinagar, J&K,
    India

R.S. Bajpai
Shri Ramswaroop Memorial
    University
Barabanki, India

Arpita Basu
Department of Electrical
    Engineering
National Institute of Technology
Jamshedpur, India

Ananyo Bhattacharya
National Institute of Technology
Jamshedpur, India

Sourav Chakraborty
Department of Electrical
    Engineering
National Institute of Technology
Rourkela, India

Chunendra Kumar Singh Chaudhary
Centre for Energy and Environment
Malaviya National Institute of
    Technology
Jaipur, Rajasthan, India

Anamika Das
National Institute of Technology
Jamshedpur, India

Kushal Jagtap
Department of Electrical Engineering
National Institute of Technology
Srinagar, Hazratbal Srinagar, J&K,
    India

Chandra Kant
Department of Mechanical Engineering
National Institute of Technology
Srinagar, J&K, India

Susmita Kar
Department of Electrical Engineering
National Institute of Technology
Rourkela, India

Dil Khush Meena
Centre for Energy and Environment
Malaviya National Institute of
    Technology
Jaipur, Rajasthan, India

B. Mouleeka
Department of Electrical Engineering
National Institute of Technology
Rourkela, India

Mamatha N.
Department of Electrical and
    Electronics Engineering University
    of Visvesvaraya College of
    Engineering
Bangalore University
Bengaluru, Karnataka, India

Ramesh H.R.
Department of Electrical and
    Electronics Engineering University of
    Visvesvaraya College of Engineering
Bangalore University
Bengaluru, Karnataka, India

Santhosha D.
Dept. of Electrical and Electronics
    Engineering University of
    Visvesvaraya College of Engineering
Bangalore University & C.I.T, Gubbi
Bengaluru, Karnataka, India

R. Sheba Rani
Department of Electrical and
    Electronics Engineering School of
    Engineering
Malla Reddy University
Hydrabad, India

Keerti Rawal
Department of Electrical Engineering
National Institute of Technology
Srinagar, J&K, India

G. Pandu Ranga Reddy
Department of Electrical and
    Electronics Engineering G. Pullaiah
    College of Engineering and
    Technology
Kurnool, India

Pradip Kumar Sadhu
Indian Institute of Technology (ISM)
Dhanbad, India

Y. Chintu Sagar
Department of Electrical and
    Electronics Engineering G. Pullaiah
    College of Engineering and
    Technology
Kurnool, India

Madhu Singh
Department of Electrical Engineering
National Institute of Technology
Jamshedpur, India

Vijay Pal Singh
Department of Electrical
    Engineering National Institute
    of Technology
Srinagar, Hazratbal Srinagar, J&K,
    India

Arti Singhal
Centre for Energy and Environment
Malaviya National Institute of
    Technology
Jaipur, Rajasthan, India

Sunanda Sinha
Centre for Energy and
    Environment
Malaviya National Institute
    of Technology
Jaipur, Rajasthan, India

Apoorva Srivastava
Babu Banarasi Das Institute of Technology
    & Management
Lucknow India

1 Comprehensive Review of Grid Operation with Distributed Resources and Charging Stations for Electric Vehicles

Santhosha D., Ramesh H.R., and Mamatha N.

1.1 INTRODUCTION: OVERVIEW OF DISTRIBUTED RESOURCES AND CHARGING STATIONS FOR ELECTRIC VEHICLES

Electric vehicles (EVs) are propelled by batteries, a form of energy storage. The fact that there are more EVs on the road presents a challenge because more of them require charging, which puts a strain on the power infrastructure [1]. There will be an increase in EVs due to the widespread manufacture of electric vehicles and affordable energy prices [2]. As they require several batteries, the energy density, weight, availability, and cost of the batteries used in electric cars are all important considerations [3, 4]. Lithium-ion batteries, which are advantageous economically, are the most commonly used batteries. Due to their great effectiveness, low weight, quick charging time, high power output, long lifespan, and low environmental impact during battery disposal, lithium-ion batteries have recently gained popularity on a global scale.

According to research, electric car charging is nonlinear and may result in harmonic distortion, phase imbalance, and an increase in transformer load with voltage swings and DC offset in the distribution system. This may be looked into using a variety of simulation platforms, such as both MATLAB and Simulink, that are employed to predict the effects of battery charging on numerous feeder systems. The power control unit delivers a changeable DC voltage to the battery during the charging of electric car batteries, and it also performs a number of filtering operations. In order to achieve the required current/voltage charging profile, the battery's associated battery management system (BMS) [5] supervises the battery's operating parameters, such as voltages and currents. The BMS also shapes charging speed.

A rise in the demand for electricity during peak hours and a reduction in reserve capacity are both caused by the expansion of EV charging stations [6]. A two-way power-flow charging station for electric vehicles that employs communication technologies to intelligently integrate it would make electric vehicle batteries useful utility assets. The hub for refuelling electric vehicles is thought to be a charging station

DOI: 10.1201/9781003311829-1

(CS). According to how heavily they rely on electricity as a fuel source, there are three primary categories of EVs [7].

a) Vehicles that use both gasoline and electricity, or hybrid electric vehicles (HEVs).
b) Plug-in hybrid electric vehicles (PHEVs), also recognised as extended-range electric vehicles (EREVs), are energised by both electricity and gasoline.
c) Battery electric vehicles (BEVs), which use batteries as the only source of power and are completely devoid of a gasoline engine, fuel tank, and exhaust pipe.

In the classifications of EVs, Hossain et al. [8] describe the various EV technologies and the futuristic methodologies which can be adapted to the EV and its charging.

Figure 1.1 represents the global market trend for EVs from 2010–2020. The majority shareholders for EV production are also in the graph. EV classifications have a huge range. There are more differences between EVs with the converters, motors, range, and strategies for using EVs efficiently. Based on classifications, most EVs are BEVs, HEVs, and PHEVs with different transmission and connection networks.

The electricity demands would also increase in the near future once EVs become popular around cities. Theis would result in more production of electricity through hybrid sources/renewable sources such as solar or wind. Vehicle-to-grid (V2G) technology would also come into play for the improvement of electrification and urbanisation.

A healthy relationship between the grid and the EV would fetch more reliability for the transmission and grid system. The exchange of effective power between the two would result in more adaption of components and easy flow of power bi-directionally.

Over the past ten years, the incorporation of renewable energy sources and EVs into distribution systems has gained significant attention due to the growing concern over the exhaustion of fossil fuels and the effects of global warming [9]. Auxiliary services including peak power reduction, voltage alteration, and improved stability are provided to distribution systems by EVs with V2G capabilities [10]. The

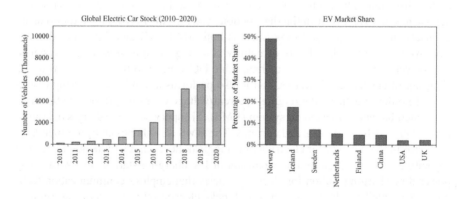

FIGURE 1.1   Global market car stock (Source: [8])

disadvantages of renewable energy's stochastic nature have been successfully mitigated, allowing the distribution system to run more efficiently and economically [11].
The synchronised allocation of distributed generating resources (DGRs) and electric
vehicle charging stations (EVCSs) in a V2G context merits exhaustive research, just
like other integration strategies for renewable energies and EVs.

The best distribution of DGRs, EVCSs, and EV V2G technologies has recently
been a hot topic in research. The following reviews the pertinent literature. Research
has been conducted from a variety of angles and has made reference to a variety of
application scenarios in relation to the best allocation of DGRs and EVCSs, as shown
in Figure 1.2.

To reduce the annual operating costs of the distribution system, DGRs like wind
turbines, solar panels, diesel generators, and energy storage devices are planned pro-
actively [12]. Power system operation constraints and EV owner requirements are
fully taken into consideration in [13] because the distribution of EVCSs appropriately
encompasses a reliability check of the electrical power system. A two-stage process
is suggested in [14] for the coordinated allocation of DGRs and EVCSs in distribution systems, during which the financial details to ensure EV parking lot investors and the restrictions of distribution system operators (DSOs) are fully taken into
consideration. To make the allocation model straightforward, the branch power-flow
constraints in [15] are laid back by a second-order conic. As a result, the utilisation
model for Photovoltaic (PV) power generation and EVCSs is constructed as a second-order cone programming (SOCP) model. This model is optimally convex and
can be successfully solved using solvers that are readily available in the marketplace. In order to expedite the time-consuming calculation required by a variety of
operation circumstances, [16] also demonstrates an accelerated generalised benders
decomposition algorithm. A Markov chain Monte Carlo (MCMC) simulation model
is employed in [17] to compensate for the uncertainty of EV charging requests and
the generation of renewable energy. Following that, the allocation model is updated
to take into account the coordinated control of energy storage charging and discharging behaviours, DGR outputs, and EV charging demands. [18] asserts that a pricing
mechanism is developed to distinguish between the V2G price and the market power

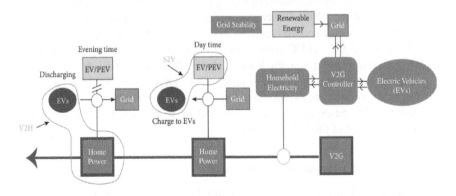

FIGURE 1.2 EV-to-grid and vice versa (Source: [8])

price, which promotes EV owners' participation in V2G activities while simultaneously assuring the financial stability of aggregators. In-depth research on combined
procurements of auxiliary amenities for EVs with V2G features is also conducted by
[19] to further expand the market environment.

1.2 SURVEY OF INFRASTRUCTURE AND CHARGING STATION CLASSIFICATIONS

Both at residences and at businesses, slow charging is a sort of charging that is used.
The specialised metering circuit in the house accepts the plug from the electric car.
Charging infrastructure can be divided into three categories based on the reliability
and availability of the grid: utility grid (UG), off-grid, and hybrid. These charging
infrastructures can also be classified as AC, DC, or hybrid AC/DC systems according to the type of power source demanded for EV charging [20, 21].

1.2.1 UG CHARGING INFRASTRUCTURE

In locations with a dependable UG, UG charging infrastructure is frequently
deployed. Here, one must plug in the EV charger connector at their residence or at an
appropriate business location, and the UG provides the AC power [22]. The AC single-phase or three-phase input can deliver this AC power to the internal and external
EV chargers. Commercial systems are widely accessible, and this infrastructure has
been well investigated and described in the literature [23]. It is not discussed in this
essay and is not the primary objective.

1.2.2 OFF-GRID CHARGING INFRASTRUCTURE

In off-grid charging infrastructure (OGCS), numerous energy sources, including fossil fuels and renewable sources, are used singly or in combination to produce and
consume electricity locally. Energy buffers (local backup) that an OGCS may have
include supercapacitors, fuel cells, and battery banks. Since it has no contact with the
UG, this system can be used in locations without it. [24] provides a diagram of the
multiple elements of such a system, including the energy sources, EV charger, Energy
Management strategy (EMS), Power electronics converters (PECs), and energy storage
system (ESS). EVs may end up receiving both AC and DC charging from an OGCS
through the use of AC and DC connectors. If implemented correctly, this framework
can also generate energy to nearby loads, including homes and individual gadgets.

According to the type of power supply (AC or DC), the OGCS can be divided
into three categories, and [24] provides a summary of relevant works. An AC-based
architecture, a DC-based architectural style, and an AC plus DC-based design are the
three distinct kinds of OGCS architecture.

1.2.3 HYBRID CHARGING INFRASTRUCTURE

According to [24] and as depicted in Figure 1.3, an interaction between UG and
OGCS results in a hybrid charging system (HCS). In other words, an OGCS can be
connected to the UG. HCSs can therefore operate in both disconnected and connected

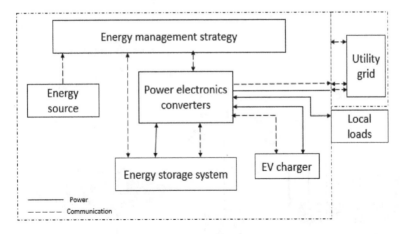

FIGURE 1.3 HCS and its interconnection

grid modes (depending on UG availability). The generation of power in this system may come from fossil fuels, the UG, or both. When there is an excess of energy generated, it is sent to the UG and then removed to keep the EVs charged when other energy sources are depleted. This advantage eliminates the need for using a sizable ESS like those found in OGCSs.

HCSs' AC-based architecture has received more attention in the literature than OGCSs' AC-based architecture [25–33]. Only PV is employed as a renewable resource in [25–33], while [30] uses only wind energy. For this charging system, the research provided in [34–50] is based on computation, tests, or a blend of the two.

A lot of research has been conducted on the DC-based HCS architecture as well as various hybrid and off-grid charging system architectures [51–53]. A bipolar DC-based architecture for HCS has also been proposed and researched in [54], consisting of two DC buses. Here, a 1.38 MW charging point simulation in MATLAB is showcased, and a scaled-down test vehicle is used to verify the suggested methodology.

The research into AC- and DC-based HCS layouts is receiving much more attention in the literature than OGCS architecture [55–59]. The fact that all of these works rely on experimental confirmation is evident here. It is stated in [58] that the maximum EV charging power is 3.7 kW.

In [60], the author proposed EV charging in Rwanda, an area in Africa with work on renewable energy sources for EVs. The plan of the paper was to increase the amount of off-grid utilisation to 52% for EVs and reduce the on-grid utilisation to the 48%. The author presented a proposed method using the HOMER grid software, where renewable energy sources such as solar power were used to connect to the microgrids for the application of EV charging.

1.3 GRID EV CHARGING TECHNOLOGY

Figure 1.4 is a representation of the author in the paper with respect to the generation, transmission, and charging of EVs using microgrids for various applications using the converters. The technology is mainly divided into V2G and grid-to-vehicle

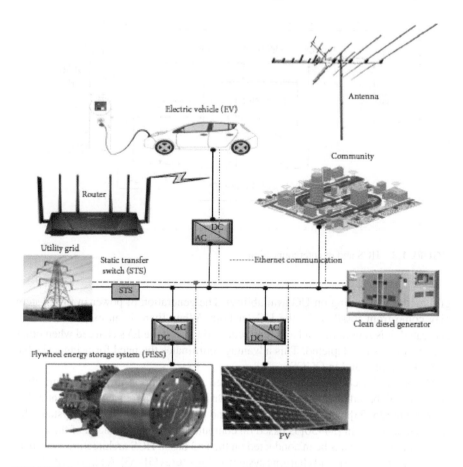

FIGURE 1.4   Grid-to-EV charging architecture (Source: [60])

(G2V). The V2G technology gives back the power to the grid utilisation, while G2V supplies power from the grid to the EV through charging stations. Smart charging technology has been proposed where the power can be made smart using the SCADA system for monitoring, and the power data can be exchanged through the grids to improve the reliability of the EV charging stations, etc. For the proposed system, the authors have chosen HOMER grid software, which can collect data on hourly daily basis for optimisation and can also perform sensitive analysis of the power exchange.

The author proposed methodology using Figure 1.5, where the current system is shown in Figure 1.5(a) and the proposed method in Figure 1.5(b). In the proposed method, the charging of the EV can be done through the UG and as well as through renewable energy sources such as solar, along with batteries. In this manner the structure would be able to utilise both the AC and DC power (through renewable energy sources) without utilisation of more AC grid power.

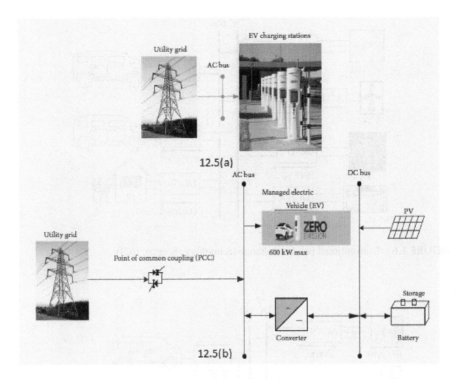

FIGURE 1.5   Grid EV technology (Source: [61])

The paper [61] presented the system planning of the grid-connected EVCS. Optimal planning of EV charging through the advanced algorithms can be used for this application to reduces carbon emissions from EVs. The author assessed 140 papers and concluded the best method for EVCSs is to use a hybrid system design with both an AC grid and renewable energy sources.

The typical system for the charging of EVs is shown Figure 1.6. Hybrid energy sources along with the AC grid and battery storage devices are used for the charging of electric vehicles and for the utilisation of power by domestic applications. The grid is converted to the DC bus system along with the wind power. These are then given to the load sections.

The proposed method in the paper is shown Figure 1.7. The system uses the renewable energy sources such as wind and solar along with the battery source to supply power to the DC bus. The DC bus also uses the AC grid power using the converter to convert AC to DC power. These are given to the load applications such as EVCSs, domestic power applications, and other loads. The added advantage in this system is the V2G technology, as explained earlier, where the excess power is transferred back from the vehicle to the grid for better utilisation of power. This system increases the efficiency and the reliability for the bus systems, as the power exchange is bi-directional.

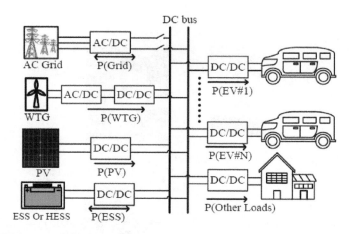

FIGURE 1.6 Conventional power exchange technology (Source: [62])

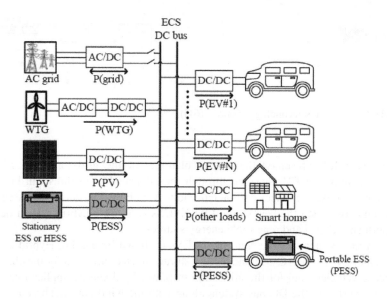

FIGURE 1.7 Proposed grid-to-EV and EV-to-grid technology [61]

1.4 DISTRIBUTED RESOURCES WITH MICROGRIDS

The operational structure for microgrids and EV charging stations has been presented in paper [62]. The paper describes the hierarchical structure, where the system is divided into the DSO, from which the data is sent to the microgrid central controller (MGCC) level, which acts as an aggregator for the data management system, as shown in Figure 1.8. Then, these [62] are divided or sent to local controller (LC)

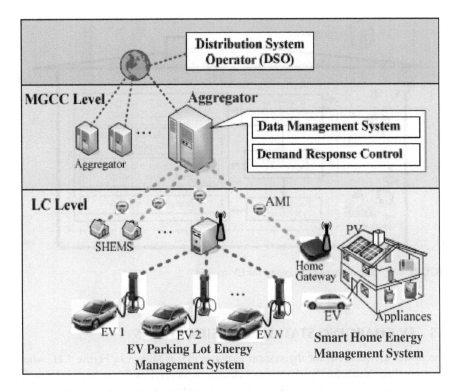

FIGURE 1.8 Different levels of power exchange (Source: [62])

levels, where the system constitutes smart home energy systems along with EV parking and charging systems.

Power data management is done through these three levels in the hierarchical model presented. The smart energy home system also acts as an important system for the V2G technology transfer. The system shown in Figure 1.9 represents the smart energy system where various algorithms with advanced methodologies can be used to detect and monitor the optimal transfer of power data for the home applications along with the EV charging and power transfer from the vehicle to the grid.

The charging infrastructure with respect to the distributed energy sources with the capability of electric and hybrid mobility is presented in paper [63]. The paper also presents the various technologies for the integration of various DC or AC sources to the grid for various applications (Figure 1.9).

Figure 1.10 represents the two architectures for V2G technology. One is the direct method, and another is the aggregative indirect method. The direct method deals with the grid to load applications without any aggregator. The power exchange cannot be determined using this method, as there are no aggregators. The second method uses aggregator technology, where the power exchange can be determined for various applications like commercial, domestic, and electric vehicle charging.

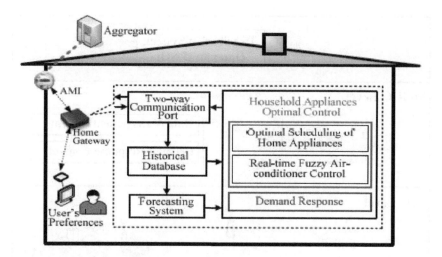

FIGURE 1.9   Smart house technology for EV-to-grid

1.5   EV CHARGING STATION CLASSIFICATIONS

The power architecture for the system in paper [63] is shown in Figure 1.11, where there are three levels for the charging of electric vehicles. Level 1 is slow charging for various converters. The second level is semi-fast charging for single-phase convert-ers, and the third level is fast charging for the three-phase converters and off-board chargers. These are used for EV and PHEV architectures.

The various technologies and methodologies for V2G and G2V have been pre-sented in [8]. The author presented an example of methodologies for charging EVs from vehicle-to-grid using the bi-directional charging station. This station would help in utilising the excess power from the vehicle to the grid so that there would be no wastage. Figure 1.12 represents the method mentioned in [8].

Figure 1.12 also depicts the sun to vehicle (S2V) infrastructure used to charge EVs using solar panels through controllers and converters to make the voltage constant at the load side. The last system shown is about the vehicle to infrastructure system, which works in a bi-directional form between the controller stations through the grid and the EV.

Paper [64] presents the EV charging infrastructure with better reliability and effi-ciency using the wired and wireless communication of the power using the grid lines. Figure 1.12 represents the detailed power system using the generation, transmission, and distribution to the various load applications along with EV charging through the charging stations at domestic and commercial sites. The stations can be used either from public places or private places.

The system can be also used for centralised and decentralised EV charging sys-tems through the hierarchical control mentioned earlier. The system can be monitored in real time with local and area controllers for the EV charging applications. This is controlled through a centralised system. In a decentralised system, an intelligent

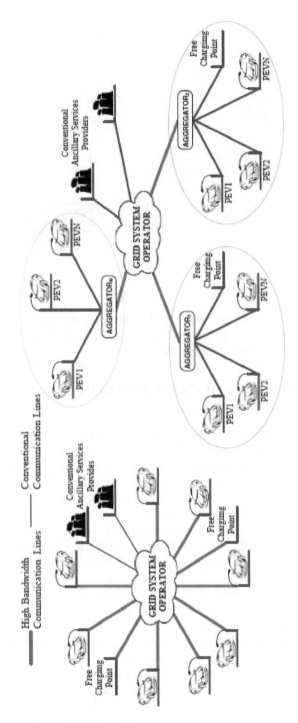

FIGURE 1.10 Vehicle-to-grid with and without aggregator technology (Source: [63])

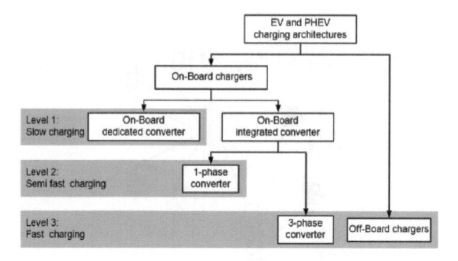

FIGURE 1.11   Charging specifications (Source: [63])

controller can be used to track and monitor all the charging stations available at some fixed location where it becomes easy to control the database of the system for EV charging, as shown in Figure 1.13 and 1.14.

1.6   DIFFERENT MODES OF GRID OPERATION AND THEIR COMPARISON

There are different modes of operation when it comes to grid technology. The technology includes islanded operation and hybrid operation. The mix of various renewable energy sources would result in hybrid technology, while the islanded technology would include the same hybrid system but somewhere in a rural area. This technology would result in better performance for the grid, as the islanded network also gives some power to the grid from various places.

Figure 1.15 represents the operation of a microgrid controller, which can be done through two methods: islanded and grid-connected [65]. The diagram represents the three-phase inverter with the supply given to the controller containing current control mode and voltage control mode. The microgrid controller plays an important role when it comes to the connection between islanded or grid-connected systems. Microgrid controllers help in achieving efficient power exchange between the portals using both the operations mentioned.

In the study presented in [66] (Figure 1.16), the grid operation for both islanded and grid-type systems using hybrid renewable energy sources is described. The energy is taken from wind and solar (PV) energy sources for hybrid energy. The battery also acts as a source for the operation. All these are connected to the microgrid, where the energy is taken and is sent to the power grid. It shows the generation from both islanded and grid-type energy sources. Various controllers and methodologies

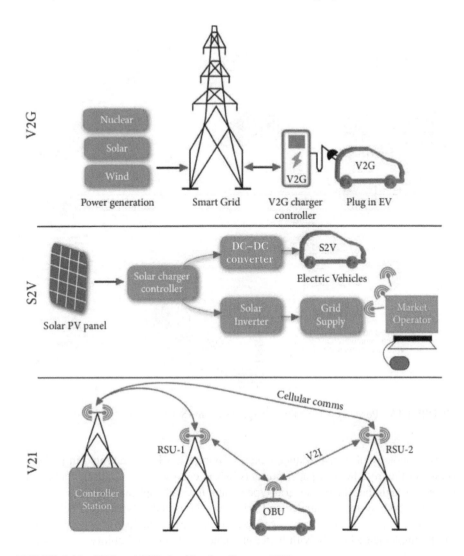

FIGURE 1.12   V2G and S2V classification (Source: [64])

are used for the efficient and reliable operation of the system. In operation, the normal mode is worked through the islanded or grid-type system, and in the case of low power, the system can be connected to the power grid or can be stopped to generate power.

Figure 1.17 represents the operation of the grid using islanded hybrid energy with enabled EV charging and applications for domestic and commercial purpose. The energy is taken from both solar and wind energy, and these power sources are monitored and controlled through the controller or the microgrid. These are then distributed from this microgrid to either the utility grid, critical and non-critical loads, or the EV charging stations.

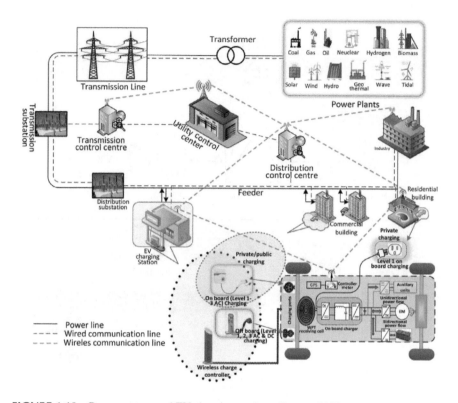

FIGURE 1.13 Power system and EV charging stations (Source: [64])

The topology of hybrid islanded energy sources has been presented in [69] and shown in Figure 1.18. The paper presents hybrid energy that is controlled and connected through the islanded or grid-type system for load applications.

Out of both systems, islanded and grid-type, the islanded type yields better performance, as there is more space available in rural areas and hybrid energy sources can be implemented easily with better efficiency and better reliability, as shown in Figure 1.19.

1.7 PROPOSED METHODOLOGY

EV charging points found on motorways, in parking areas, and in malls have high rapid charging current rates of 30–70 A in comparison to home charging, which utilises 12–16 A. Consumer satisfaction is increased by the brief charging time, which also offers charging flexibility. The distribution network experiences less peak demand when EV charging stations are in the ideal location.

The demand on the grid will undoubtedly rise due to the rapid growth of EVs, which will have an impact on the generation system. It would be expensive to build sufficient electricity-generating facilities to accommodate this demand. However, merging the existing model with renewable energy sources like rooftop solar, wind

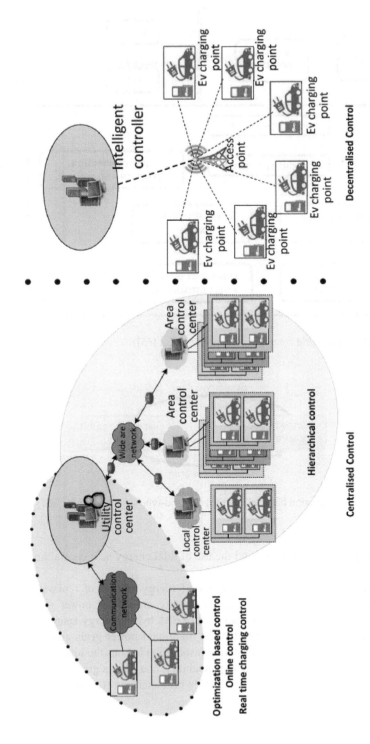

FIGURE 1.14 Centralised and decentralised methods of EV charging (Source: [64])

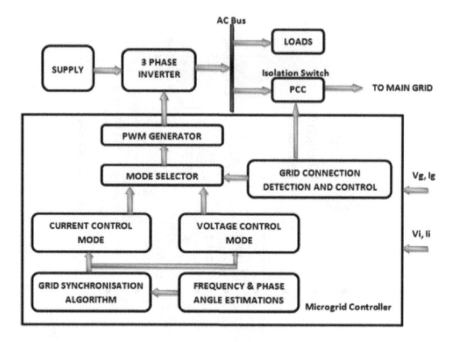

FIGURE 1.15   Microgrid controller operation (Source: [65])

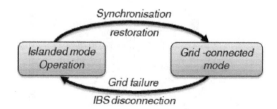

FIGURE 1.16   Connection between islanded and grid-type systems (Source: [67])

power, energy storage batteries, and the use of tidal energy near coastal areas is one option.

The intermittent nature and battery-based energy storage of renewable energy sources are drawbacks. Because they can easily meet EVs' power requirements with the aid of renewable energy resources (RER), hybrid energy sources allow EV charging stations to be less dependent on traditional power grids and hence more independent. Hybrid EV charging station installation lessens the load on the electrical grid, increases charging station dependability, and lowers greenhouse gas emissions [4]. In Figure 1.20, the suggested EV hybrid charging station block diagram is displayed.

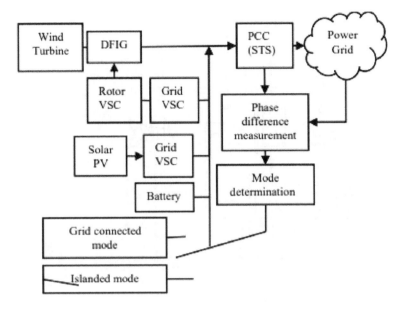

FIGURE 1.17 Islanded and grid-connected operation (Source: [66])

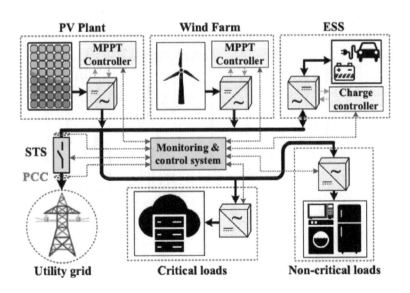

FIGURE 1.18 Whole operation representation of islanded hybrid system (Source: [68])

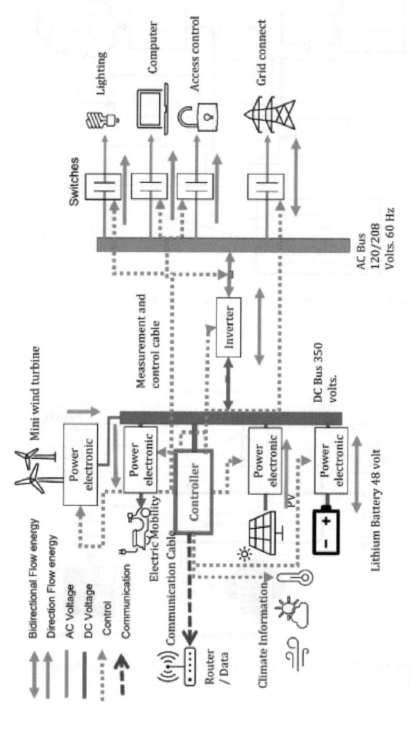

FIGURE 1.19 Islanded type hybrid energy system (Source: [69])

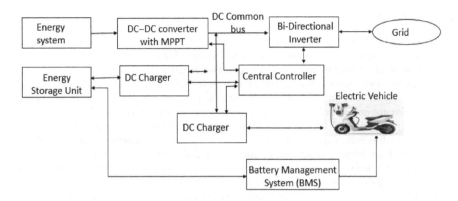

FIGURE 1.20 EV hybrid charging system

1.8 CONCLUSION

The effects of electric vehicle charging stations on the electrical grid are covered in this study along with the findings of relevant studies. Depending on how many and how far electric vehicles can travel, charging stations for them must be available. The demand for alternative energy sources is growing in today's rapidly evolving technological environment. The transportation sector's contribution to lowering greenhouse gas emissions and combating global warming is significant. It is impossible to ignore how significantly the widespread integration of electric vehicle charging stations into the current grid network impacts the grid characteristics.

The incorporation of hybrid technology in EV charging stations, which combines solar, wind, and other renewable energy sources with battery storage systems, may eventually improve the stability and availability of the power supply and lessen its impact on the electrical grid. Planning and construction of EV charging stations in the future should evaluate the implications of those stations on the local power grid. Future EV charging station planning and construction should take into consideration how charging stations will affect the associated power grid.

The analysis also reveals that while four-wheeler charging power levels and connectors vary from nation to nation, they are generally standardised. One-, two-, and three-wheelers still use unusual charging connections on the vehicle side, making up a sizeable portion of the EV market. This can be a problem for EV manufacturers who operate on a worldwide scale. As a result, it is recommended to employ a variety of solutions, such as limiting the quantity of USB, micro-B, and USB type-C ports on the vehicle's side or implementing wireless charging for these EVs.

ACKNOWLEDGEMENT

Our sincere gratitude and appreciation go out to our professor and the entire department for their invaluable counsel and support.

REFERENCES

1. Sanchari Deb, Karuna Kalita, and Pinakeshwar Mahanta, "Review of impact of electric vehicle charging station on the power grid," in *2017 IEEE International Conference on Technological Advancements in Power and Energy*, 2017.
2. Sanchari Deb, Karuna Kalita, and Pinakeshwar Mahanta, "Impact of electric vehicle charging stations on reliability of distribution network," in *2017 IEEE International Conference on Technological Advancements in Power and Energy*, 2017.
3. A. Aljanad and Azah Mohamed, "Harmonic impact of plug-in hybrid electric vehicle on electric distribution system," *Modeling and Simulation in Engineering*, vol. 2016, 2016.
4. Abdul Rauf Bhatti, Zainal Salam, Mohd Junaidi Bin Abdul Aziz, and Kong Pui Yee, "A critical review of electric vehicle charging using solar photovoltaic," *International Journal of Energy Research*, vol. 40, pp. 439–461, 2016.
5. Electric vehicles in India and its impact on Grid, https://www.nsgm.gov.in/sites/default/files/EV-in-India-and-its-Impact-on-Grid-May-2017.pdf.
6. R. Pawełek, P. Kelm and I. Wasiak, "Experimental analysis of DC electric vehicles charging station operation and its impact on the supplying grid," *2014 IEEE International Electric Vehicle Conference (IEVC)*, Florence, Italy, 2014, pp. 1–4.
7. www.ergon.com.au/network/smarter-energy/electric-vehicles/types-of-electric-vehicles.
8. M. S. Hossain, Laveet Kumar, Mamdouh El Haj Assad, and Reza Alayi, "Advancements and future prospects of electric vehicle technologies: A comprehensive review," *Complexity, Hindawi*, vol. 2022, pp. 1–21, July 2022.
9. L. Mehigan, J. P. Deane, B. P. Ó. Gallachóir, and V. Bertsch, "A review of the role of distributed generation (DG) in future electricity systems," *Energy*, vol. 163, pp. 822–836, 2018.
10. C. Guille and G. Gross, "A conceptual framework for the vehicle-to-grid (V2G) implementation," *Energy Policy*, vol. 37, no. 11, pp. 4379–4390, 2009.
11. S. Aghajani and M. Kalantar, "Optimal scheduling of distributed energy resources in smart grids: A complementarity approach," *Energy*, vol. 141, pp. 2135–2144, 2017.
12. S. Mahdavi, R. Hemmati, M. A. Jirdehi, "Two-level planning for coordination of energy storage systems and wind-solar-diesel units in active distribution networks," *Energy*, vol. 151, pp. 954–965, 2018.
13. S. Davidov and M. Pantoš, "Optimization model for charging infrastructure planning with electric power system reliability check," *Energy*, vol. 166, pp. 886–894, 2019.
14. M. H. Amini, M. P. Moghaddam, and O. Karabasoglu, "Simultaneous allocation of electric vehicles' parking lots and distributed renewable resources in smart power distribution networks," *Sustainable Cities and Society*, vol. 28, pp. 332–342, 2017.
15. H. Zhang, S. Moura, Z. Hu, W. Qi, and Y. Song, "Joint PEV charging station and distributed PV generation planning," in *IEEE Power and Energy Society General Meeting*, 2017.
16. H. Zhang, S. J. Moura, Z. Hu, W. Qi, and Y. Song, "Joint PEV charging network and distributed PV generation planning based on accelerated generalized benders decomposition," *IEEE Transactions on Transportation Electrification*, vol. 4, no. 3, pp. 789–803, 2018.
17. S. M. Kandil, H. E. Z. Farag, M. F. Shaaban, and M. Z. El-Sharafy, "A combined resource allocation framework for PEVs charging stations, renewable energy resources and distributed energy storage systems," *Energy*, vol. 143, pp. 961–972, 2018.
18. T. Mao, W. Lau, C. Shum, H. S. Chung, K. Tsang, and N. C. Tse, "A regulation policy of EV discharging price for demand scheduling," *IEEE Transactions on Power Systems*, vol. 33, no. 2, pp. 1275–1288, 2018.
19. S. Faddel, A. Aldeek, A. T. Al-Awami, E. Sortomme, and Z. Al-Hamouz, "Ancillary services bidding for uncertain bidirectional V2G using fuzzy linear programming," *Energy*, vol. 160, pp. 986–995, 2018.

20. M. C. Catalbas, M. Yildirim, A. Gulten and H. Kurum, "Estimation of optimal locations for electric vehicle charging stations," *2017 IEEE International Conference on Environment and Electrical Engineering and 2017 IEEE Industrial and Commercial Power Systems Europe (EEEIC/I&CPS Europe)*, Milan, Italy, 2017, pp. 1–4.

21. L. Luo, Z. Wu, W. Gu, H. Huang, S. Gao, and J. Han, "Coordinated allocation of distributed generation resources and electric vehicle charging stations in distribution systems with vehicle-to-grid interaction," *Energy*, vol. 192, p. 116631, 2020.

22. M. R. Khalid, M. S. Alam, A. Sarwar, and M. Jamil Asghar, "A comprehensive review on electric vehicles charging infrastructures and their impacts on power-quality of the utility grid," *eTransport*, vol. 1, 2019, Art. no. 100006.

23. T. U. Solanke, V. K. Ramachandaramurthy, J. Y. Yong, J. Pasupuleti, P. Kasinathan, and A. Rajagopalan, "A review of strategic charging discharging control of grid-connected electric vehicles," *Journal of Energy Storage*, vol. 28, 2020, Art. no. 101193.

24. G. Rituraj, G. R. C. Mouli, and P. Bauer, "A comprehensive review on off-grid and hybrid charging systems for electric vehicles," *IEEE Open Journal of the Industrial Electronics Society*, vol. 3, pp. 203–222, 2022, doi: 10.1109/OJIES.2022.3167948.

25. A. Verma and B. Singh, "Energy management strategy of solar PV-battery and diesel generator based electric vehicle charging station," in *IEEE Energy Conversion Congress and Exposition*, pp. 1043–1050, 2018.

26. H. Zhao and A. Burke, "An intelligent solar powered battery buffered EV charging station with solar electricity forecasting and EV charging load projection functions," in *Proceedings of IEEE International Electric Vehicle Conference*, pp. 1–7, 2014.

27. D. Li, A. Zouma, J.-T. Liao, and H.-T. Yang, "An energy management strategy with renewable energy and energy storage system for a large electric vehicle charging station," *eTransport*, vol. 6, 2020, Art. no. 100076.

28. U. Datta, A. Kalam, and J. Shi, "Smart control of Bess in PV integrated EV charging station for reducing transformer overloading and providing battery-to-grid service," *Journal of Energy Storage*, vol. 28, 2020, Art. no. 101224.

29. Y. Yang, Q.-S. Jia, G. Deconinck, X. Guan, Z. Qiu, and Z. Hu, "Distributed coordination of EV charging with renewable energy in a microgrid of buildings," *IEEE Transactions on Smart Grid*, vol. 9, no. 6, pp. 6253–6264, November 2018.

30. B. Ye, J. Jiang, L. Miao, P. Yang, J. Li, and B. Shen, "Feasibility study of a solar-powered electric vehicle charging station model," *Energies*, vol. 8, no. 11, pp. 13265–13283, 2015.

31. A. Verma and B. Singh, "Multimode operation of solar PV array, grid, battery and diesel generator set based EV charging station," *IEEE Transactions on Industry Applications*, vol. 56, no. 5, pp. 5330–5339, September/October 2020.

32. S. M. Shariff, M. S. Alam, F. Ahmad, Y. Rafat, M. S. J. Asghar, and S. Khan, "System design and realization of a solar-powered electric vehicle charging station," *IEEE Systems Journal*, vol. 14, no. 2, pp. 2748–2758, June 2020.

33. R. Wang, Q. Sun, D. Qin, Y. Li, X. Li, and P. Wang, "Steady-state stability assessment of ac-busbar plug-in electric vehicle charging station with photovoltaic," *Journal of Modern Power Systems and Clean Energy*, vol. 8, no. 5, pp. 884–894, 2020.

34. A. Mathapati, P. M. Mouna Machamma, B. N. Rangaswamy, and R. Sandeep Kumar, "Impact of connecting electric vehicles to grid and its challenges," *JETIR*, vol. 6, no. 5, May 2019.

35. O. Marcincin, Z. Medvec, and P. Moldrik, "The impact of electric vehicles on distribution network," in *2017 18th International Scientific Conference on Electric Power Engineering (EPE)*, Kouty nad Desnou, Czech Republic, 2017, pp. 1–5.

36. Y. Huang, J. Liu, X. Shen, and T. Dai, "The interaction between the large-scale EVs and the power grid," *Smart Grid and Renewable Energy*, vol. 4, no. 2, pp. 137–143, 2013, doi: 10.4236/sgre.2013.42017.

37. S. Painuli, M. S. Rawat, and D. Rao Rayudu, "A comprehensive review on electric vehicles operation, development and grid stability," in *2018 International Conference on Power Energy, Environment and Intelligent Control (PEEIC)*, pp. 807–814, 2018, doi: 10.1109/PEEIC.2018.8665643.

38. S. Bimenyimana, C. Wang, A. Nduwamungu, G. N. Osarumwense Asemota, W. Utetiwabo, C. L. Ho, and J. De Dieu Niyonteze, "Integration of microgrids and electric vehicle technologies in the national grid as the key enabler to the sustainable development for Rwanda," *International Journal of Photoenergy*, vol. 2021, 2021, Art. ID 9928551, https://doi.org/10.1155/2021/9928551.

39. A. Jenn and J. Highleyman, "Distribution grid impacts of electric vehicles: A California case study," *iScience*, vol. 25, no. 1, p. 103686, 2022.

40. W. Samantha, "Share of energy related carbon dioxide emissions in China in 2019," www.statista.com/statistics/1088662/china-share-of-energy-related-carbondioxide-emissions-by-sector/.

41. EPA. [Online]. Available: www.epa.gov/ghgemissions/sources-greenhouse-gas-emissions. [Accessed: 30 October 2022].

42. "Greenhouse gas emissions from transport in Europe," *European Environment Agency*, 25 October 2021. [Online]. Available: www.eea.europa.eu/data-and-maps/indicators/transport-emissions-of-greenhouse-gases-7. [Accessed: 30 October 2022].

43. NIESJ, "Japan's national greenhouse gas emissions," 2020, www.env.go.jp/press/814.pdf.

44. NIESJ, "Japan's national greenhouse gas emissions," in *National Institute for Environmental Studies, Japan, Fiscal Year 2019 (Final Figures)*, 2021, nies.Go.jp/whats-new/20210413/20210413-e.html.

45. A. H. Akinlabi and D. Solyali, "Configuration, design, and optimization of air-cooled battery thermal management system for electric vehicles: A review," *Renewable and Sustainable Energy Reviews*, vol. 125, 2020, Art. ID 109815.

46. A. Sharma and S. Sharma, "Review of power electronics in vehicle-to-grid systems," *Journal of Energy Storage*, vol. 21, pp. 337–361, 2019.

47. J. Wang, J. Huang, and M. Dunford, "Rethinking the utility of public bicycles: The development and challenges of station-less bike sharing in China," *Sustainability*, vol. 11, no. 6, p. 1539, 2019.

48. S. Hussain et al., "The emerging energy internet: Architecture, benefits, challenges, and future prospects," *Electronics*, vol. 8, no. 9, p. 1037, 2019.

49. L. A. Alwal, P. K. Kihato, and S. I. Kamau, "A review of control strategies for microgrid with PV-wind hybrid generation systems," in *Proceedings of the Sustainable Research and Innovation Conference*, 2018.

50. G. Dhananjay and R. P. Hema, "Microgrid system advanced control in islanded and grid connected mode," in *2014 IEEE International Conference on Advanced Communication Control and Computing Technologies (ICACCCT)*, pp. 301–305, 2014.

51. B. Sun, "A multi-objective optimization model for fast electric vehicle charging stations with wind, PV power and energy storage," *Journal of Cleaner Production*, vol. 288, 2021, Art. no. 125564.

52. G. R. C. Mouli, P. Bauer, and M. Zeman, "Comparison of system architecture and converter topology for a solar powered electric vehicle charging station," in *Proceedings of 9th International Conference on Power Electronics, ECCE Asia*, pp. 1908–1915, 2015.

53. S. A. Singh, N. A. Azeez, and S. S. Williamson, "A new single-stage high-efficiency photovoltaic (PV)/grid-interconnected dc charging system for transportation electrification," in *Proceedings of 44th Annual Conference of the IEEE Industrial Electronics Society*, pp. 005374–005380, 2015.

54. S. Rivera and B. Wu, "Electric vehicle charging station with an energy storage stage for split-DC bus voltage balancing," *IEEE Transactions on Power Electronics*, vol. 32, no. 3, pp. 2376–2386, March 2017.

55. A. Verma and B. Singh, "A solar PV, BES, grid and DG set based hybrid charging station for uninterruptible charging at minimized charging cost," in *Proceedings of IEEE Industry Applications Society Annual Meeting*, pp. 1–8, 2018.

56. B. Singh and A. Verma, "Control of solar PV and WEGS powered EV charging station," in *Proceedings of IEEE Transportation Electrification Conference*, pp. 609–614, 2020.

57. B. Singh, A. Verma, A. Chandra, and K. Al-Haddad, "Implementation of solar PV-battery and diesel generator based electric vehicle charging station," *IEEE Transactions on Industry Applications*, vol. 56, no. 4, pp. 4007–4016, July/August 2020.

58. G. R. Chandra Mouli, P. Van Duijsen, F. Grazian, A. Jamodkar, P. Bauer, and O. Isabella, "Sustainable e-bike charging station that enables AC, DC and wireless charging from solar energy," *Energies*, vol. 13, no. 14, 2020, Art. no. 3549.

59. A. Verma and B. Singh, "CAPSA based control for power quality correction in PV array integrated EVCS operating in standalone and grid connected modes," *IEEE Transactions on Industry Applications*, vol. 57, no. 2, pp. 1789–1800, March/April 2021.

60. C.-T. Ma, "System planning of grid-connected electric vehicle charging stations and key technologies: A review," *Energies*, vol. 12, no. 21, p. 4201, 2019.

61. A. Verma and B. Singh, "CAPSA based control for power quality correction in PV array integrated EVCS operating in standalone and grid connected modes," *IEEE Transactions on Industry Applications*, vol. 57, no. 2, pp. 1789–1800, March/April 2021.

62. J. T. Liao, C. I. Lin, C. Y. Chien, and H. T. Yang, "The distributed energy resources operation for EV charging stations and SHEMS in microgrids," in *2014 International Conference on Intelligent Green Building and Smart Grid (IGBSG)*, pp. 1–6, 2014, doi: 10.1109/IGBSG.2014.6835256.

63. Clemente Capasso, Sebastian Riviera, Samir Kouro, and Ottorino Veneri, "Charging architectures integrated with distributed energy resources for sustainable mobility," *Energy Procedia*, vol. 105, pp. 2317–2322, 2017, ISSN 1876–6102, https://doi.org/10.1016/j.egypro.2017.03.666.

64. H. S. Das, M. M. Rahman, S. Li, and C. W. Tan, "Electric vehicles standards, charging infrastructure, and impact on grid integration: A technological review," *Renewable and Sustainable Energy Reviews*, vol. 120, p. 109618, 2020, ISSN 1364–0321, https://doi.org/10.1016/j.rser.2019.109618.

65. M. Megha Prakash and Jasmy Paul, "Control of microgrid for different modes of operation," *International Journal of Engineering Research*, vol. 5, no. 5, 2016.

66. Zhao Dongmei, Zhang Nan, and Liu Yanhua, "Micro-grid connected/islanding operation based on wind and PV hybrid power system," pp. 1–6, 2012, 10.1109/ISGT-Asia.2012.6303168.

67. A. Bidram, V. Nasirian, A. Davoudi, and F. L. Lewis, "Cooperative synchronization in distributed microgrid control," *Advances in Industrial Control*, vol. 16, pp. 242–249, 2017, Springer International Publishing.

68. J. C. Vasquez, J. M. Guerrero, A. Luna, P. Rodriguez, and R. Teodorescu, "Adaptive droop control applied to voltage-source inverters operating in grid-connected and islanded modes," *IEEE Transactions on Industrial Electronics*, vol. 56, no. 10, pp. 4088–4096, October 2009, doi: 10.1109/TIE.2009.2027921.

69. D. Akinyele, J. Belikov, and Y. Levron, "Challenges of microgrids in remote communities: A STEEP model application," *Energies-MDPI*, vol. 11, no. 432, February 2018.

2 A Comprehensive Study of Life Cycle Assessments of Electric Vehicles and IC Engine Vehicles

Chunendra Kumar Singh Chaudhary, Dil Khush Meena, Arti Singhal, and Sunanda Sinha

2.1 INTRODUCTION

The demand for automobiles in the transportation sector is increasing on a daily basis due to the significant rise in global population growth [1]. The automobile sector, or road transportation, contributes a significant portion of global emissions in the category of transportation [2]. In 2021, overall CO_2 emissions from the transportation sector increased by 8% to nearly 7.7 Gt CO_2, up from 7.1 Gt CO_2 in 2020, with the automobile sector accounting for 76.60% of emissions (5.86 Gt CO_2) [3]. The most popular vehicles in the automobile sector are ICEVs (petrol/diesel), but ICEVs run on fossil fuels, which have a considerable negative impact on the environment, causing climatic changes, environmental pollution, and energy shortages [4]. All of these issues have encouraged researchers and manufacturers to develop innovative technology in the automotive sector.

However, EVs have emerged as the best option among all best alternatives for vehicles that combine various multiple technologies, such as HEVs, FCEVs, and PHEVs, and emit fewer air pollutants than ICEVs in the transportation industry [5–7]. As a result, many nations are embarking on an electrical transition in order to expand the use of EVs due to their short- and long-term prospects for a more sustainable environment and sustainable growth to overcome this scenario [8]. According to IEA data, 10 million EVs were produced globally before 2021 (Figure 2.1). As of right now, China is the world's leading producer of EVs, followed by the United States, Germany, the United Kingdom, France, Norway, Korea, Italy, the Netherlands, and Sweden (Figure 2.2). In the Sustainable Development Scenario, the number of EVs worldwide is projected to reach over 70 million in 2025 and 230 million in 2030 [4].

However, there are many disagreements about how environmentally friendly electric vehicles are in the different phases of their lives [9]. Reviewing the numerous environmental impacts of Evs in comparison to those of ICEVs is therefore vital for a

DOI: 10.1201/9781003311829-2

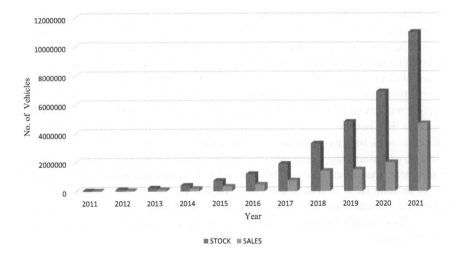

FIGURE 2.1   Global EV sales and stock (2011–2021)

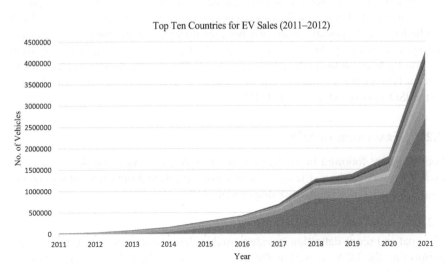

FIGURE 2.2   Top 10 countries for EV sales worldwide (2011–2021)

sustainable environment. The life phases (production, use, and EOL), electricity use scenario, battery production and degradation, and material use are the different phases and aspects for which the environmental consequences of Evs can be calculated [10].

Compared to other vehicles with conventional engines, EVs have a higher environmental impact during the manufacturing phase than they do during the use phase. In addition, they have a smaller overall life-cycle impact on GWP and a greater overall life-cycle impact on HTP [10]. And as can be seen from the electricity use scenario, the environmental impact of EVs is largely dependent on the electricity that will be used during the use phase of the vehicle. Using non-renewable sources of electricity has a greater negative impact on the environment than using renewable sources of electricity [11]. Regarding the EOL phase, repurposing and remanufacturing dead batteries, can improve the environmental advantages of EVs throughout the recycling process [10]. LCA is the best approach for quantifying those environmental impacts of Evs because it can assess a variety of environmental impacts, including GHG emissions, toxicity potential, mineral depletion, and land use [12].

In order to assess the environmental implications of EVs for a more sustainable environment, the author analysed numerous LCA studies that have been conducted on EVs in contrast to ICEVs in this book chapter. Further sections are arranged as Section 2.2, which is the LCA methodology section, Section 2.3, a comparison of LCAs of EVs and ICEVs at various stages of their lives, and Section 2.4, the conclusion of the study.

2.2 METHODOLOGY OF LCA

LCA is defined as the process used to evaluate a product, process, or activity's potential environmental consequences throughout the course of its life cycle [13]. LCA can help to clarify possible trade-offs between different environmental impacts and life cycle stages [14]. And the wide scope of LCA is useful for transferring from one life-cycle phase to another, one region to another, or one environmental issue to another [15]. General standards and procedures to evaluate LCA have been defined by the ISO 14,040 and ISO 14,044 [16].

2.2.1 FRAMEWORK OF LCA

Four steps are followed in the framework of LCA: goal and scope definitions, life cycle inventory analysis, life cycle impact assessment, and life cycle interpretation. Figure 2.3 shows the framework diagram of LCA.

2.2.1.1 Goal and Scope Definitions

The goal and scope definition is widely used to describe the reason and purpose for performing the LCA, as well as the intent to conduct the LCA, and describes the major functional units, quantity and quality of data, and system requirements.

2.2.1.2 Life Cycle Inventory Analysis

Inventory analysis is used to compile a list of environmental inputs and outputs to perform the LCA; in this phase, the data from the environment may be research data, practical data, or both.

2.2.1.3 Life Cycle Impact Assessment

The primary objective of life cycle impact assessment is to analyse the environmental consequences of the product in the context of the complete LCA based on the results of the life cycle inventory analysis phase. This is the most essential

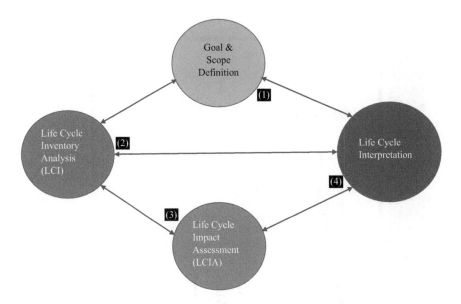

FIGURE 2.3   Basic framework of LCA

part of the LCA since it offers precise information on the product's environmental impacts.

2.2.1.4   Life Cycle Interpretation

After analysing the data from the previous three phases, this final phase of LCA is used to discuss the results and significant issues that have come up in the product LCA and to evaluate the results in order to reduce the negative environmental effects of the product via sensitivity analysis and consistency checks.

2.2.2   CLASSIFICATION OF LCA

The overall LCA is mainly classified into three types: cradle-to-gate, cradle-to-grave, and cradle-to-cradle, having three subcategories: well-to-tank, tank-to-wheel, and well-to-wheel, as shown in Figure 2.4 [13].

2.2.2.1   Cradle-to-Gate

Cradle-to-gate LCA is described as the environmental assessment of a product or service from its manufacturing phase to the gate phase before the product will go into its use phase. As seen from all phases of life, the use phase and recycling phase are not considered in this type of LCA. [17].

2.2.2.2   Cradle-to-Grave

Cradle-to-grave LCA is an evaluation of a product's environmental impact from its manufacturing phase to its EOL phase and covers the extraction of raw materials,

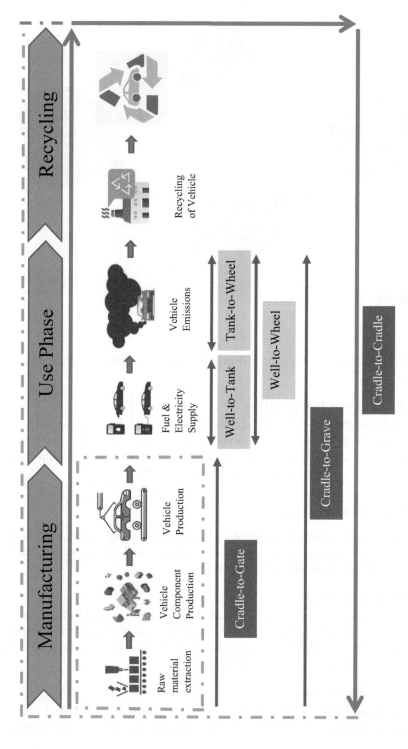

FIGURE 2.4 Classification of LCA considering different phases (Source: [13])

component and vehicle production, the use phase, and the recycling and dismantling of vehicle components [13].

2.2.2.3 Cradle-to-Cradle

Cradle-to-cradle is also named closed-loop LCA. In this type of LCA, the construction waste and recycled products are reused at the processing stage of raw material. This helps in reducing the environmental impact of products and encouraging manufacturers to shift towards more sustainable production methods, as well as incorporating social responsibility into product development [13].

Three subcategories of LCA exist: well-to-tank, tank-to-wheel, and well-to-wheel. Well-to-tank LCA focuses on the energy source to the vehicle's tank storage, the focus of tank-to-wheel LCA is mainly on the energy carrier and fuel tank for the vehicle's propulsion system during operation, and well-to-wheel LCA focuses on the life cycle of the energy source, such as gasoline, utilised to move the vehicle [18].

2.2.3 LCA Software Tools

It is difficult to estimate all of the environmental impacts at once. Regarding this, there are numerous software programmes developed to calculate these environmental impacts and perform an LCA. But the selection of an appropriate software tool that can be used in LCA is crucial, as is the work involved, as each has its own unique characteristics, such as database availability, functionality, data quality management, user interface, and modelling principles for developing product systems. Thus, the databases utilised in the computation and technique are essential components of these technologies. As these two elements are interdependent, the majority of available software is suited for working with one or more specific databases [19]. The various LCA tools and software programmes available on the market are shown in Table 2.1.

As the majority of LCA software tools are now available on the market, they can be purchased or downloaded for no or little cost [20]. Among the all software programmes, the best are CMLCA, Simapro, Umberto, GEMIS, OpenLCA, Mobius, and GaBi, as shown in Figure 2.5 [13].

SimaPro, GaBi, Open LCA, and Umberto were the most popular and widely used software tools because they provide a variety of features. Among them, as seen from the last 15 years of market data, GaBi and Simapro are the most frequently used tools in the research work [21].

2.2.4 LCA Impact Parameters

Various LCA parameters by which comparison or environmental assessment has been checked are GWP, HTP, EP, ADP, AP, and PM shown in Figure 2.6.

2.3 LCA COMPARISON OF EVs AND ICEVs

ICEVs and EVs are the most common vehicles in the category of vehicles. ICEVs are vehicles that use only an ICE to meet their energy needs and are classified as

TABLE 2.1
Different software programmes based on LCA

S.No.	Software/Tool	Webpage	Developers of Tool
1.	SIMAPRO	www.pre-sustainability.com	PRé-Consultants
2.	GaBi	www.gabi-software.com	Pe-International
3.	UMBERTO	www.umberto.de/en	ifu Hamburg GMBH
4.	OpenLCA	www.openlca.org	GreenDelta GmbH
5.	SABENTO	www.sabento.com	ifu Hamburg GmbH, Germany
6.	Ecochain Mobius	https://ecochain.com/mobius	Ecochain Technologies B.V.H.J.E. Wenckebachweg, Amsterdam, Netherlands
7.	GEMIS	www.gemis.de	Oeko Institute, Germany
8.	TESPI	www.elca.enea.it	ENEA, Italy
9.	LEGEP	www.legep.de/?lang=en	LEGEP Software GmbH, Germany
10.	EARTHSTER 2 TURBO	www.greendelta.com	GreenDelta GmbH
11.	SULCA 4.2	www.pre-sustainability.com	VTT Technical Research Centre (Finland)
12.	E^3 DATABASE	www.e3database.com	Ludwig-BölkowSystemtechnik GmbH, Germany
13.	ECO-BAT 4.0	www.eco-bat.ch	Haute Ecole d'Ingénierie et de Gestion du Canton de Vaud, Switzerland
14.	TEAM	www.ecobilan.pwc.fr/en/boite-a-outils/team.jhtml	Ecobilan PricewaterhouseCoopers
15.	REGIS	www.sinum.com	Sinum AG—EcoPerformance Systems
16.	USES-LCA 2.0	www.cem-nl.eu/useslca.html	Netherlands Center For Environmental Modeling
17.	AIST-LCA 4	www.aist-riss.jp/main	National Institute of Advanced Industrial Science and Technology (AIST), Japan
18.	CMLCA 4.2	www.cml.leiden.edu/software	Leiden University, Institute of Environmental Sciences (CML), Holland
19.	BEES 4.0	www.nist.gov/el/economics/BEESSoftware.cfm	National Institute of Standards and Technology, USA
20.	Ecochain Mobius Athena	https://ecochain.com/mobius www.athenasmi.org	Ecochain Technologies B.V. H.J.E. Wenckebachweg, Amsterdam, Netherlands Athena Sustainable Materials Institute, Canada

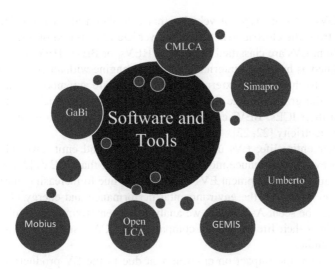

FIGURE 2.5   Suitable Software's of LCA

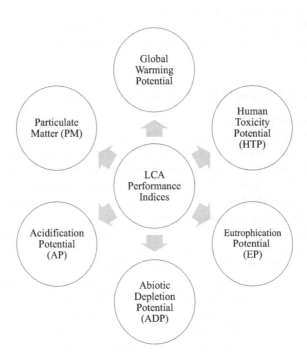

FIGURE 2.6   Various impact parameters of LCA. The considered environmental parameters for estimating the performance of EVs over their life cycle are global warming potential (GWP), human toxicity potential (HTP), eutrophication potential (EP), abiotic depletion potential (ADP), acidification potential (AP), and particulate matter (PM)

GVs and DVs. EVs are type of vehicles that can obtain all or a portion of their electricity from the electric grid to run the vehicle [22]. Based on their degree of electrification, EVs are classified as HEVs, PHEVs, or BEVs. HEVs are a category of EVs defined as having an internal combustion engine with an electric motor but no capability to charge from external source. PHEVs are automobiles that use electricity from the grid to charge their batteries and alternative fuels (petrol, diesel, etc.) to run their ICEs. BEVs entirely depend on batteries, with no ICE, and run entirely on electricity [22, 23].

Over their entire life, EVs use fewer fossil fuels and emit fewer GHGs than ICEVs; however, they produce more EP, HTP, and PM than ICEVs [24, 25]. In the past few years, the development EVs has increased due to increasing energy crises and pollution. However, the environmental performance and energy efficiency of EVs have also be seen. As a result, we analysed the environmental implications of EVs throughout their life cycles and compared the LCA results of EVs and ICEVs in multiple studies.

In conclusion, the impact on environment due to the EV production phase is increased due to the manufacturing of batteries in this phase, but as seen from the other phases, the use of clean energy and battery recycling make EVs perform better; therefore, impacts on environment during the use and recycling stages will be reduces. EVs use less fossil fuel and emit fewer GHGs than ICEVs over their entire life cycle, but in the manufacturing of batteries, the consumption of mineral and metal resources is so high that EVs have more impact on ADP and HTP [26, 13].

2.3.1 LCA OF EVs AND ICEVs IN VARIOUS LIFE PHASES

The evaluation comprises the manufacture, use, and EOL or recycling stages of EVs and ICEVs in order to define their total life cycle implications [9]. The manufacturing or production process includes the construction process, from extraction of raw material to the manufacture of vehicle components and vehicle manufacturing, in contrast to the use phase, which encompasses energy generation and emissions during operation. In the EOL or recycling phase, the remaining material has been recycled utilising various recycling techniques [27].

Due to battery production, EVs have a greater environmental impact than ICEVs during the production or manufacturing phase; however, EVs have a lower environmental impact than ICEVs during the use phase [28–30].

The resources used in all phases of an EV's life also have a significant role in determining its environmental impact. In the production category, the material type for component and vehicle production is critical. In the use phase, the category of fuel or electricity used must be considered. In the EOL phase and in the recycling phase, the material type for recycling plays a significant role [26].

2.3.1.1 Manufacturing Phase

Materials extraction, parts manufacturing, vehicle assembly, and vehicle distribution are the stages covered in the manufacturing or production phase [31]. During the manufacturing phase of a bus, car, or motorcycle, the following components' emissions were calculated: (1) chassis of ICEVs; (2) engine and gearbox; (3) motor, inverter, and battery for EVs [32].

During the production process, EVs emit between 14.6 and 14.7 tonnes of CO_2, which is 59–60% more than ICEVs (9.2 tonnes) [33]. In comparison to ICEVs, the production phase environmental consequences are higher for EVs due to battery manufacture [34, 35].

2.3.1.2   Use Phase

The use phase contains two sub-stages of automobile operation: energy generation and emissions while driving. Energy production includes all transformation processes that occur prior to fuel consumption, including fuel generation from feedstock production, transportation, feedstock conversion to final fuel, and subsequent storage, distribution, and vehicle tank delivery. The impacts on the energy supply chain are computed using resource depletion and emissions associated with the generation of fuel for ICEVs and electricity for BEVs used during operation [27].

The electricity used to charge EVs has a substantial effect on the environmental impact of the vehicle, such that using clean power grids and renewable energy sources can help to lower the GHG emissions of EVs during their use phase, and this can compensate for the emissions that occur in production phase [36–41].

2.3.1.3   EOL or Recycling Phase

The EOL phase refers to the 2000/53/EC Directive and ISO 22628:2002 "Road Vehicles Recyclability and Recoverability: Calculation Method" (ISO 22628, 2002) [42], which divide vehicle EOL into four different steps: depollution, disassembly, shredding, and post-shredding. The environmental impact EVs in the EOL phase can be determined by calculating the energy spent by the disassembly/recycling/landfill processes, the credits obtained by recyclable material and energy flows, and the environmental emissions created by rubbish landfilling. Recycling EV batteries can enhance resource utilisation and reduce GHG and SO_2 emissions significantly [27].

During the recycling process, vehicles must be disassembled before the metal and non-metallic parts of each component are separated and purified. Following the preceding step, some of the vehicle's raw materials can be recycled. Due to the presence of heavy metal electrolytes and other contaminants in EV power batteries, the energy consumption and carbon emissions associated with the recycling phase of a vehicle are quantified separately for battery recycling and non-battery recycling. Metals are recycled for non-battery parts, while non-metallic elements, such as plastic and glass, are dumped in landfills or burned as waste, according to the recycling method commonly employed by Chinese domestic dismantling firms [43].

Because the battery is an important component of an EV but EOL battery parts have a number of environmental consequences, there is a need to recycle the battery parts that remain in EOL EVs. The techniques that can be used to recycle battery parts are pyrometallurgy, hydrometallurgy, direct regeneration, and bioleaching [10].

2.3.2   LCA Results of EVs and ICEVs in Several Studies

Because each researcher utilises a different set of data for analysis, a different piece of software, or a different database source, comparing the results of multiple authors reveals that the LCA outcomes are distinct for each [10, 13]. Technical features, such as the weight of the car without passengers and luggage, as well as battery weight, are significant factors in determining the environmental emissions during the production phase while analysing the LCA of BEVs and ICEVs. The following crucial aspects during the use phase to think about are the type of fuel and the consumption level of fuel.

Additionally, the date of manufacture affects the consumption level of fuel, as the newest model consumes less fuel than the previous model. The power sources used for battery charging in EVs also have a closer impact on environment, with thermal power plants and major hydropower plants having a greater impact on the environment than wind power plants and solar PV power plants. As a broad comparison, researchers use a region's electrical composition to evaluate the total impact of charging in that area. As the generation of electricity shifts from fossil fuels to renewable resources, energy production will become more environmentally friendly.

The comparison of LCA results achieved by various authors for various types of EVs and ICEVs is summarised in Table 2.2.

2.4   CONCLUSION AND RECOMMENDATIONS

Due to the rise in vehicle numbers brought on by population expansion, the automotive industry contributes significantly to CO_2 and air pollution emissions. Electrifying automobiles is a key step in addressing the aforementioned problems. Because of how batteries affect the environment, EVs may not be environmentally friendly; hence, this chapter compared EVs to regular ICEVs in terms of how their life-cycle environmental impacts compared.

Due to the substantial volumes of chemicals and metals that are required for the production of batteries, EVs have a greater influence on HTP, EP, ADP, AP, and PM than ICEVs. When it comes to use, EVs have much better environmental performance than ICEVs, although this also greatly relies on the type of electricity utilised, with clean and renewable energy having a lesser impact on the environment. The recycling, reuse, and remanufacturing of batteries fall under the category of the EOL phase and may have negative effects on the environment during the EV recycling stage. Environmental advantages during the use and recycling phases of EVs can undoubtedly offer a way to make up for their environmental drawbacks in the manufacturing phase. To make EVs more environment friendly, it is thought that using a mix of electricity sources as well as improvements in battery manufacturing and recycling technologies will be essential.

In order to promote and develop EVs in the long term, it is crucial to use clean energy sources or more renewable energy sources, improve battery manufacturing technologies and battery categories, and actively develop a number of methods for the recycling, remanufacturing, and reuse of batteries as the number of EVs on the market increases in near future.

TABLE 2.2
LCA results of EVs and ICEVs in several studies

Reference	Location and Time Framework	Vehicle type	Focus of the study	Results and Conclusion
[44]	USA, China, France, South Korea, Germany, Sweden, Norway, Netherlands, Canada, and the UK (2019–2030)	EVs	➢ Cradle-to-grave LCA of EVs using Simapro9.1 software for 10 selected countries. ➢ Taking electricity mix scenario of (2019–2030).	➢ 2019: China (>42,000 kg CO_2 eq.), Sweden (13,209 kg CO_2 eq.) ➢ 2025: South Korea (35,471 kg CO_2 eq.), Sweden (13,297 kg CO_2 eq.) ➢ 2030: South Korea (35,179 kg CO_2 eq.), Norway (13,251 kg CO_2 eq.)
[45]	Hong Kong (2019–2050)	BEVs, PHEVs, ICEVs	➢ LCA comparison of EVs, ICEVS, and PHEVs using Simapro software. ➢ Mixed scenario of electricity generation (2019–2050).	➢ In five of the eight impact categories that were chosen, including GWP (15,712 kg CO_2 eq.), FPM (26 kg $PM_{2.5}$ eq.), FD (4,285 kg oil eq.), TE (55,121 kg 1,4-DB eq.), and OZF (71 kg NO_x eq.) the environmental impact is lower with the 2050 electricity mix scenario.
[46]	Spain (2018–2050)	BEVs, HEVs, ICEVs (Petrol & Diesel)	➢ To assess and compare the environmental impacts of EVs, HEVs, and ICEVs, cradle-to-grave LCA is used. ➢ Mixed electricity generation scenario (2018–2050).	➢ BEVs might potentially cut 27% of their present GWP by 2050 as compared to the current scenario of 2018. According to the scenario discussed, BEVs are capable of eliminating 48% of gasoline ICEV GHG emissions.
[47]	China, Germany, USA, Italy (2020)	EVs, ICEVs	➢ Estimating the effects of EVs and ICEVs on the environment. ➢ Considering four alternative situations, each assuming different nations where the various stages of a vehicle's life cycle occur.	➢ Comparing the same vehicle segment under the same conditions, EVs always have lower CO_2 emissions than ICEVs. ➢ Variations in the specific energy requirements for battery production, vehicle use, and energy-related CO_2 emission levels have a substantial effect on the overall CO_2 emissions of an electric car.

(Continued)

TABLE 2.2 (Continued)
LCA results of EVs and ICEVs in several studies

Reference	Location and Time Framework	Vehicle type	Focus of the study	Results and Conclusion
[31]	China (2020)	BEVs, PHEVs, ICEVs	➢ Assessment of CO_2 emissions and air pollutant in different stages (production, use, EOL) ➢ Comparison of emissions of ICEV, PHEV, BEV.	➢ Comparing a BEV to an ICEV results in reductions of 9.7 kg in VOCs, 2.2 kg in NO_x, and 6.2 t of CO_2, and but increases in SO_2 of 28.5 kg and in $PM_{2.5}$ of 4.0 kg. ➢ In comparison to the ICEV, the PHEV reduced emissions by 6.7 kg of VOCs, 1.2 kg of NO_x, and 1.4 t of CO_2, but increased SO_2 by 14.2 kg and $PM_{2.5}$ by 1.9 kg per vehicle.
[48]	Lithuania (2015–2050)	BEVs, ICEVs (petrol and diesel)	➢ Comparative cradle-to-grave LCA of EVs and ICEVs. ➢ Different electricity mix scenarios (2015–2050).	➢ BEVs powered by the electricity mix of 2015 emit 47% and 26% more GHGs than ICEVs powered by diesel and gasoline, respectively. ➢ The impact of BEVs with the 2050 electricity mix is 54% less than that of BEVs with the 2015 electricity mix.
[49]	Hebei Province, China (2015–2030)	BEVs, ICEVs	➢ Cradle-to-grave LCA of BEVs and ICEVs.	➢ Compared to ICEVs, promoting BEVs might lower the consumption of fossil fuel by 25–50% and use per-kilometre of petroleum by 98%. ➢ BEVs have clear benefits over ICEVs in terms of reducing NO_x, CO, VOCs, $PM_{2.5}$, and CO_2 emissions, but their PM_{10} emissions are greater. ➢ The conclusion shows that increasing clean electricity can improve the environmental impacts of EVs.

Ref.	Location	Vehicles	Focus	Findings
[5]	Czech Republic and Poland (2015–2050)	EVs, ICEVs	➤ Comparative cradle-to-grave LCA of EVs and ICEVs using Simapro. ➤ Considering future electricity mix scenario (2015–2050).	➤ Comparing EVs with ICEVs found that GHG emissions and FD will be lower for EVs in the Czech Republic and Poland as seen from both present and future scenarios. ➤ EVs have more effect on TA, FE, HTP, and FPM than ICEVs. ➤ An increase in renewable electricity can decrease the overall emissions of BEVs.
[50]	Lithuania (2015–2050)	BEVs, PHEVs, ICEVs	➤ Comparative environmental cradle-to-grave LCA of BEVs, PHEVs, and ICEVs. ➤ Mixed electricity scenario (2015–2050).	➤ In all the categories, petrol ICEVs cause higher environmental impacts. ➤ The BEV with a 2050 electricity mix has 33%, 43%, and 27% less impact on the environment than the BEV (2015 electricity mix), petrol ICEV, and diesel ICEV, respectively.
[51]	China, United States (2012–2050)	EVs	➤ Evaluation of GHG emissions and air pollutants (SO_2, PM_{10}, $PM_{2.5}$, and NO_x) of EVs.	➤ GHGs and air pollutant reductions of 60–85% could be achieved if EVs are charged from clean electricity sources.

REFERENCES

1. V. Nimesh, R. Kumari, N. Soni, A. K. Goswami, and V. Mahendra Reddy, "Implication viability assessment of electric vehicles for different regions: An approach of life cycle assessment considering exergy analysis and battery degradation," *Energy Convers Manage*, vol. 237, Jun. 2021, doi: 10.1016/J.ENCONMAN.2021.114104.
2. B. Marmiroli, M. Venditti, G. Dotelli, and E. Spessa, "The transport of goods in the urban environment: A comparative life cycle assessment of electric, compressed natural gas and diesel light-duty vehicles," *Appl Energy*, vol. 260, Feb. 2020, doi: 10.1016/J.APENERGY.2019.114236.
3. "Transport—topics—IEA," www.iea.org/topics/transport (accessed Sep. 28, 2022).
4. "Global EV outlook 2022—analysis—IEA," www.iea.org/reports/global-ev-outlook-2022 (accessed Sep. 26, 2022).
5. D. Burchart-Korol, S. Jursova, P. Folęga, J. Korol, P. Pustejovska, and A. Blaut, "Environmental life cycle assessment of electric vehicles in Poland and the Czech Republic," *J Clean Prod*, vol. 202, pp. 476–487, Nov. 2018, doi: 10.1016/J.JCLEPRO.2018.08.145.
6. T. R. Hawkins, B. Singh, G. Majeau-Bettez, and A. H. Strømman, "Comparative environmental life cycle assessment of conventional and electric vehicles," *J Ind Ecol*, vol. 17, no. 1, pp. 53–64, 2013, doi: 10.1111/J.1530-9290.2012.00532.X.
7. A. Nordelöf, M. Romare, and J. Tivander, "Life cycle assessment of city buses powered by electricity, hydrogenated vegetable oil or diesel," *Transp Res D Transp Environ*, vol. 75, pp. 211–222, Oct. 2019, doi: 10.1016/J.TRD.2019.08.019.
8. G. Kamiya, J. Axsen, and C. Crawford, "Modeling the GHG emissions intensity of plug-in electric vehicles using short-term and long-term perspectives," *Transp Res D Transp Environ*, vol. 69, pp. 209–223, Apr. 2019, doi: 10.1016/J.TRD.2019.01.027.
9. C. Tagliaferri et al., "Life cycle assessment of future electric and hybrid vehicles: A cradle-to-grave systems engineering approach," *Chem Eng Res Des*, vol. 112, pp. 298–309, Aug. 2016, doi: 10.1016/J.CHERD.2016.07.003.
10. X. Xia and P. Li, "A review of the life cycle assessment of electric vehicles: Considering the influence of batteries," *Sci Total Environ*, vol. 814, p. 152870, Mar. 2022, doi: 10.1016/J.SCITOTENV.2021.152870.
11. Ö. Andersson and P. Börjesson, "The greenhouse gas emissions of an electrified vehicle combined with renewable fuels: Life cycle assessment and policy implications," *Appl Energy*, vol. 289, p. 116621, May 2021, doi: 10.1016/J.APENERGY.2021.116621.
12. E. Helmers, J. Dietz, and M. Weiss, "Sensitivity analysis in the life-cycle assessment of electric vs. combustion engine cars under approximate real-world conditions," *Sustainability (Switzerland)*, vol. 12, no. 3, Feb. 2020, doi: 10.3390/su12031241.
13. S. Verma, G. Dwivedi, and P. Verma, "Life cycle assessment of electric vehicles in comparison to combustion engine vehicles: A review," *Mater Today Proc*, vol. 49, pp. 217–222, 2021, doi: 10.1016/J.MATPR.2021.01.666.
14. R. Garcia and F. Freire, "A review of fleet-based life-cycle approaches focusing on energy and environmental impacts of vehicles," *Renew Sust Energy Rev*, vol. 79, pp. 935–945, 2017, Elsevier Ltd, doi: 10.1016/j.rser.2017.05.145.
15. R. Turconi, A. Boldrin, and T. Astrup, "Life cycle assessment (LCA) of electricity generation technologies: Overview, comparability and limitations," *Renew Sust Energy Rev*, vol. 28, pp. 555–565, 2013, Elsevier Ltd, doi: 10.1016/j.rser.2013.08.013.
16. "ISO—ISO 14044:2006—Environmental management—life cycle assessment—requirements and guidelines," www.iso.org/standard/38498.html (accessed Sep. 28, 2022).
17. M. Messagie et al., "Life cycle assessment of conventional and alternative small passenger vehicles in Belgium," *2010 IEEE Vehicle Power and Propulsion Conference, VPPC 2010*, 2010, doi: 10.1109/VPPC.2010.5729233.

18. L. Gao and Z. C. Winfield, "Life cycle assessment of environmental and economic impacts of advanced vehicles," *Energies (Basel)*, vol. 5, no. 3, pp. 605–620, 2012, doi: 10.3390/en5030605.
19. M. Ormazabal, C. Jaca, and R. Puga-Leal, "Analysis and comparison of life cycle assessment and carbon footprint software," *Adv Intell Syst Comput*, vol. 281, pp. 1521–1530, 2014, doi: 10.1007/978-3-642-55122-2_131/COVER.
20. D. Aparecido Lopes Silva, V. Aparecida da Silva Moris, C. Moro Piekarski, U. Diogo Aparecido Lopes Silva, A. Oliveira Nunes, and T. Oliveira Rodrigues, "How important is the LCA software tool you choose comparative results from GaBi, openLCA, SimaPro and Umberto," *researchgate.net*, 2017, (accessed: Sep. 26, 2022). [Online]. Available: www.researchgate.net/profile/Thiago-Rodrigues-20/publication/318217178_ How_important_is_the_LCA_software_tool_you_choose_Comparative_results_ from_GaBi_openLCA_SimaPro_and_Umberto/links/5fbe51c0299bf104cf75bc02/ How-important-is-the-LCA-software-tool-you-choose-Comparative-results-from-Ga- Bi-openLCA-SimaPro-and-Umberto.pdf.
21. D. A. Lopes Silva, A. O. Nunes, C. M. Piekarski, V. A. da Silva Moris, L. S. M. de Souza, and T. O. Rodrigues, "Why using different life cycle assessment software tools can generate different results for the same product system? A cause–effect analysis of the problem," *Sustain Prod Consum*, vol. 20, pp. 304–315, Oct. 2019, doi: 10.1016/J. SPC.2019.07.005.
22. H. Ma, F. Balthasar, N. Tait, X. Riera-Palou, and A. Harrison, "A new comparison between the life cycle greenhouse gas emissions of battery electric vehicles and internal combustion vehicles," *Energy Policy*, vol. 44, pp. 160–173, May 2012, doi: 10.1016/j. enpol.2012.01.034.
23. A. Nordelöf, M. Messagie, A. M. Tillman, M. Ljunggren Söderman, and J. van Mierlo, "Environmental impacts of hybrid, plug-in hybrid, and battery electric vehicles—what can we learn from life cycle assessment?" *Int J Life Cycle Assess*, vol. 19, no. 11, pp. 1866–1890, Oct. 2014, doi: 10.1007/S11367-014-0788-0.
24. D. Burchart-Korol, S. Jursova, P. Folęga, and P. Pustejovska, "Life cycle impact assessment of electric vehicle battery charging in European Union countries," *J Clean Prod*, vol. 257, Jun. 2020, doi: 10.1016/j.jclepro.2020.120476.
25. Q. Qiao et al., "Life cycle cost and GHG emission benefits of electric vehicles in China," *Transp Res D Transp Environ*, vol. 86, Sep. 2020, doi: 10.1016/J.TRD.2020.102418.
26. I. Dolganova, A. Rödl, V. Bach, M. Kaltschmitt, and M. Finkbeiner, "A review of life cycle assessment studies of electric vehicles with a focus on resource use," *Resources*, vol. 9, no. 3, Mar. 2020, doi: 10.3390/RESOURCES9030032.
27. F. del Pero, M. Delogu, and M. Pierini, "Life cycle assessment in the automotive sector: A comparative case study of Internal Combustion Engine (ICE) and electric car," *Procedia Struct Integr*, vol. 12, pp. 521–537, 2018, doi: 10.1016/J.PROSTR.2018.11.066.
28. A. Lajunen and T. Lipman, "Lifecycle cost assessment and carbon dioxide emissions of diesel, natural gas, hybrid electric, fuel cell hybrid and electric transit buses," *Energy*, vol. 106, pp. 329–342, Jul. 2016, doi: 10.1016/J.ENERGY.2016.03.075.
29. Q. Qiao, F. Zhao, Z. Liu, S. Jiang, and H. Hao, "Cradle-to-gate greenhouse gas emissions of battery electric and internal combustion engine vehicles in China," *Appl Energy*, vol. 204, pp. 1399–1411, Oct. 2017, doi: 10.1016/J.APENERGY.2017.05.041.
30. A. Yu, Y. Wei, W. Chen, N. Peng, and L. Peng, "Life cycle environmental impacts and carbon emissions: A case study of electric and gasoline vehicles in China," *Transp Res D Transp Environ*, vol. 65, pp. 409–420, Dec. 2018, doi: 10.1016/J.TRD.2018.09.009.
31. L. Yang, B. Yu, B. Yang, H. Chen, G. Malima, and Y. M. Wei, "Life cycle environmental assessment of electric and internal combustion engine vehicles in China," *J Clean Prod*, vol. 285, Feb. 2021, doi: 10.1016/j.jclepro.2020.124899.

32. S. Kosai et al., "Estimation of greenhouse gas emissions of petrol, biodiesel and battery electric vehicles in Malaysia based on life cycle approach," *Sustainability*, vol. 14, no. 10, p. 5783, May 2022, doi: 10.3390/SU14105783.

33. Q. Qiao, F. Zhao, Z. Liu, S. Jiang, and H. Hao, "Comparative study on life cycle CO_2 emissions from the production of electric and conventional vehicles in China," *Energy Procedia*, vol. 105, pp. 3584–3595, 2017, doi: 10.1016/J.EGYPRO.2017.03.827.

34. H. C. Kim, T. J. Wallington, R. Arsenault, C. Bae, S. Ahn, and J. Lee, "Cradle-to-gate emissions from a commercial electric vehicle Li-Ion battery: A comparative analysis," *Environ Sci Technol*, vol. 50, no. 14, pp. 7715–7722, Jul. 2016, doi: 10.1021/ACS.EST.6B00830.

35. P. Girardi, A. Gargiulo, and P. C. Brambilla, "A comparative LCA of an electric vehicle and an internal combustion engine vehicle using the appropriate power mix: The Italian case study," *Int J Life Cycle Assess*, vol. 20, no. 8, pp. 1127–1142, Aug. 2015, doi: 10.1007/S11367-015-0903-X.

36. D. Wang et al., "Life cycle analysis of internal combustion engine, electric and fuel cell vehicles for China," *Energy*, vol. 59, pp. 402–412, Sep. 2013, doi: 10.1016/J.ENERGY.2013.07.035.

37. R. Faria, P. Marques, P. Moura, F. Freire, J. Delgado, and A. T. de Almeida, "Impact of the electricity mix and use profile in the life-cycle assessment of electric vehicles," *Renew Sust Energy Rev*, vol. 24, pp. 271–287, Aug. 2013, doi: 10.1016/J.RSER.2013.03.063.

38. G. Zhou, X. Ou, and X. Zhang, "Development of electric vehicles use in China: A study from the perspective of life-cycle energy consumption and greenhouse gas emissions," *Energy Policy*, vol. 59, pp. 875–884, Aug. 2013, doi: 10.1016/J.ENPOL.2013.04.057.

39. Q. Qiao, F. Zhao, Z. Liu, X. He, and H. Hao, "Life cycle greenhouse gas emissions of electric vehicles in China: Combining the vehicle cycle and fuel cycle," *Energy*, vol. 177, pp. 222–233, Jun. 2019, doi: 10.1016/J.ENERGY.2019.04.080.

40. J. Li and B. Yang, "Analysis of greenhouse gas emissions from electric vehicle considering electric energy structure, climate and power economy of EV: A China case," *Atmos Pollut Res*, vol. 11, no. 6, pp. 1–11, Jun. 2020, doi: 10.1016/J.APR.2020.02.019.

41. M. Kannangara, F. Bensebaa, and M. Vasudev, "An adaptable life cycle greenhouse gas emissions assessment framework for electric, hybrid, fuel cell and conventional vehicles: Effect of electricity mix, mileage, battery capacity and battery chemistry in the context of Canada," *J Clean Prod*, vol. 317, p. 128394, Oct. 2021, doi: 10.1016/J.JCLEPRO.2021.128394.

42. "ISO—ISO 22628:2002—Road vehicles—recyclability and recoverability—calculation method," www.iso.org/standard/35061.html (accessed Sep. 28, 2022).

43. W. Li, H. Bai, J. Yin, and H. Xu, "Life cycle assessment of end-of-life vehicle recycling processes in China—Take Corolla taxis for example," *J Clean Prod*, vol. 117, pp. 176–187, Mar. 2016, doi: 10.1016/j.jclepro.2016.01.025.

44. M. Shafique and X. Luo, "Environmental life cycle assessment of battery electric vehicles from the current and future energy mix perspective," *J Environ Manage*, vol. 303, p. 114050, Feb. 2022, doi: 10.1016/J.JENVMAN.2021.114050.

45. M. Shafique, A. Azam, M. Rafiq, and X. Luo, "Life cycle assessment of electric vehicles and internal combustion engine vehicles: A case study of Hong Kong," *Res Transp Econ*, vol. 91, Mar. 2022, doi: 10.1016/J.RETREC.2021.101112.

46. G. Puig-Samper Naranjo, D. Bolonio, M. F. Ortega, and M. J. García-Martínez, "Comparative life cycle assessment of conventional, electric and hybrid passenger vehicles in Spain," *J Clean Prod*, vol. 291, p. 125883, Apr. 2021, doi: 10.1016/J.JCLEPRO.2021.125883.

47. S. Franzò and A. Nasca, "The environmental impact of electric vehicles: A novel life cycle-based evaluation framework and its applications to multi-country scenarios," *J Clean Prod*, vol. 315, p. 128005, Sep. 2021, doi: 10.1016/J.JCLEPRO.2021.128005.

48. K. Petrauskienė, M. Skvarnavičiūtė, and J. Dvarionienė, "Comparative environmental life cycle assessment of electric and conventional vehicles in Lithuania," *J Clean Prod*, vol. 246, Feb. 2020, doi: 10.1016/J.JCLEPRO.2019.119042.
49. S. Shi, H. Zhang, W. Yang, Q. Zhang, and X. Wang, "A life-cycle assessment of battery electric and internal combustion engine vehicles: A case in Hebei Province, China," *J Clean Prod*, vol. 228, pp. 606–618, Aug. 2019, doi: 10.1016/J.JCLEPRO.2019.04.301.
50. K. Petrauskienė, A. Galinis, D. Kliaugaitė, and J. Dvarionienė, "Comparative environmental life cycle and cost assessment of electric, hybrid, and conventional vehicles in Lithuania," *Sustainability*, vol. 13, no. 2, p. 957, Jan. 2021, doi: 10.3390/SU13020957.
51. H. Huo, H. Cai, Q. Zhang, F. Liu, and K. He, "Life-cycle assessment of greenhouse gas and air emissions of electric vehicles: A comparison between China and the U.S.," *Atmos Environ*, vol. 108, pp. 107–116, May 2015, doi: 10.1016/j.atmosenv.2015.02.073.

3 Electric Vehicle Charging Infrastructure

Optimal Location Problem Modeling Options and Solution Techniques

Dil Khush Meena, Arti Singhal, Chunendra Kumar Singh Chaudhary, and Sunanda Sinha

3.1 INTRODUCTION

About 15% of worldwide greenhouse gas (GHG) emissions come from the automobile industry, and this share is likely to rise in the future [1]. Governments have promoted the use of "electric" vehicles to reduce GHG emissions. Electric vehicles have the potential to cut carbon dioxide emissions almost by 28% by 2030, according to a recent study [2]. When combined with low-carbon energy generation systems, EVs have a lower global warming potential than internal combustion engine (ICE) cars during their whole lifetime. Furthermore, EVs offer a variety of advantages over ICE cars, including being significantly quieter and emitting no exhaust emissions, which may help reduce air pollution and exposure to nitrogen oxides, volatile organic compounds, and carbon monoxide in urban areas. Despite these advantages, the main barriers to EV adoption are the high price of acquiring EVs comparative to ICEs and their limited driving range [3].

In certain nations, the total cost of ownership for an EV throughout its full life cycle is already cheaper than that of an ICE car. Most EVs have a shorter driving range than ICE cars. Even if a fully charged battery's range is enough for most users' everyday travel, they are anxious that battery will run out of charge before completing their journey or reaching a charging outlet. The latest EV models now offer better driving ranges of approximately 200 miles, but the lack of suitable charging infrastructure in many parts of the world continues to impede the EV market's growth. The way ahead might be to either enhance battery capacity or to build an efficient charging infrastructure to serve the charging demands of entire territories. However, the installation of EV chargers for different charging levels is prohibitively expensive and faces several technological and economic restrictions. Therefore, optimal location planning is required for the deployment of electric vehicle charging stations (EVCSs).

It is necessary to conduct research regarding highly accurate and proper deployment techniques to place and scale EVCSs while minimizing the wastage of resources

DOI: 10.1201/9781003311829-3

so that charging network operators can meet as much charging demand as possible while staying within their investment budgets. In the first section of this chapter, the author outlines the need for solutions to optimal charging location problems. The charging infrastructure, followed by the technical, economic, and environmental concerns involved with its deployment, are discussed in Section 3.2. The problem formulation for the location optimization issue is based on the objective functions and constraints discussed in Sections 3.3 to 3.5. Section 3.6 then discusses the various methods used to address optimization problems. In addition, Sections 3.7 and 3.8 describe the techniques and algorithms utilized in optimization methods for optimization issues. Section 3.9 details the influence of EVs on the distribution system, the environment, and the economy. The summary of the chapter and the future scope are discussed in Sections 3.10 and 3.11, respectively.

3.2 ELECTRIC VEHICLE CHARGING INFRASTRUCTURE

The following framework provides context for the topic of the implementation of EV charging infrastructure: electric vehicle drivers with limited autonomy utilize the road system. These trips use energy, which depletes the EV battery's charge and necessitates recharging, which may be accomplished in one of the two ways: at home or at work, or via public (or semi-public) charging infrastructure. For operators with limited investment capacity, this infrastructure must be inexpensive while still providing EV consumers with the essential transportation network coverage. Developing a charging infrastructure for a variety of charging systems and charge levels necessitates consideration of three sorts of concerns: technical, economic, and user-centered challenges.

3.2.1 CHARGING SYSTEMS AND LEVELS

This section gives an overview of the various charging levels and charging schemes, as well as a few of the international standards. Charging levels are categorized internationally by organizations like the Electric Power Research Institute (EPRI), the Society of Automotive Engineers (SAE), and the International Electro technical Commission (IEC). As shown in Figure 3.1 there are two main types of charging systems: AC and DC [4]. Other classifications are based on type, location, and flow of power between EVs and CSs.

3.2.1.1 AC Charging System

Converting AC into DC makes battery charging easier. According to the SAE EV and IEC standards, this charging infrastructure may be classified into three distinct types, which are as follows [5].

3.2.1.1.1 Level 1 Charging

These systems charge the batteries using a nominal 120 V, 12/16 A single-phase AC and has a 2 kW power output [6]. There is no need for specialized installation of onboard chargers because these chargers effectively use household plugs. This level typically

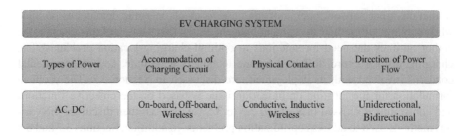

FIGURE 3.1   Classification of different types of EV charging systems

uses specific connectors (like SAEJ1772 and NEMA 5–15). However, thermal and cable size constraints result in a significantly long charging time, between 6 and 16 hours.

3.2.1.1.2   Level 2 Charging

Basic specifications for a single-phase connection are 240 V, 80 A, and around 19 kW. Typically, this technique employs the utilization of an onboard charger. Different connection types used by these chargers include SAE, Scame, and IEC-Mennekes. Level 2 AC chargers are used in most public CSs [7].

3.2.1.1.3   Level 3 Charging

These chargers utilize industry-standard connections for EVs, such as IEC, magnetic charging, and IEC-Mennekes, to supply power in three phases at 400 volts with a total output of more than 20 kilowatts. With quick chargers in this category, electric vehicle battery packs may be fully charged in 30 minutes. However, level 3 AC chargers can't be manufactured until the IEC standard is finalized. Level 3 AC power is defined as off-board charging in the SAEJ1772 standard [8].

3.2.1.2   DC Charging System

Modern EV battery charging systems utilize DC charging systems, which transform DC voltages into levels more suitable for EV battery packs. There are three types of charging level in DC charging systems, as in AC systems. For clarity, a comparative analysis of charging levels is depicted in Table 3.1.

3.2.2   EV Connectors

The connection of the EV power supply equipment (EVSE) commonly incorporates these chargers through CHAdeMO, SAE Combo, and IEC-Mennekes Combo [9]. EV connections should be made to fit the outlet on the car and the outlet on the EVCS. The market offers two types of EV connectors: AC and DC EV connectors [10]. The types of connectors with standardization, organization, and charging capability are shown in Figure 3.2.

Fast chargers are major contributors to electric distribution network issues since they consume a large quantity of power in a short length of time. Technical, economic, and environmental concerns are all broad categories that describe these problems.

TABLE 3.1

Comparative analysis of charging levels

Type	Charging Type	Input Voltage, Current	Charging Time	Power Rating (kW)	Charging Standard
Level 1 (AC)	Onboard	110 V,15 A	10–13 Hr.	1.4–1.9 kW	SAE, IEC
Level 2 (AC)	Onboard	240 V, 40 A	1–3 Hr.	19.2 kW	SAE, IEC
Level 3 (AC)	Off-board	400–500 V, 80 A	0.5–1.4 Hr.	>20 kW	SAE
Level 1 (DC)	Off-board	200–450 V, 80 A	1.2 Hr.	36 kW	SAE, CHAdeMO
Level 2 (DC)	Off-board	200–450 V, 200 A	0.4 Hr.	90 kW	SAE, CHAdeMO
Level 3 (DC)	Off-board	200–600 V, 400 A	0.2 Hr.	240 kW	SAE, CHAdeMO

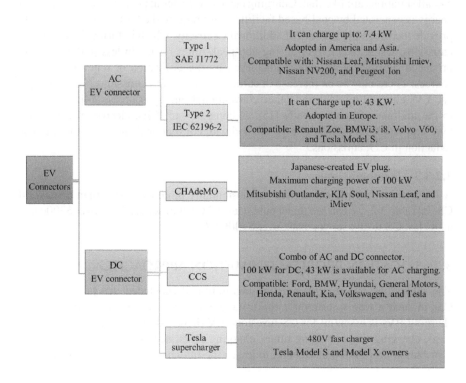

FIGURE 3.2 Different classes of charging connectors

3.2.3 Issues with EVCS Infrastructure

3.2.3.1 Technical Issues

The rapid global expansion of the EV market has introduced a new challenge for the infrastructure of distribution networks and the companies that operate them. The power consumption of a level 1 facility is about 3 kW, which is significantly less than that of a typical home. The grid should not be significantly impacted. However, the power requirements of level 3 chargers can go up to 100 kW, and the grid may not be able to support a charging hub's worth of fast chargers at once. The incorporation of excessively high-power chargers into the distribution network may result in detrimental effect on bus voltages, power efficiency, and overall system stability. The situation has the potential to cause power loss and harmonic distortion in the grid [11]. If we want to keep the price of infrastructure in check, we need to make this decisions based on the actual needs of customers, the capacity of the electricity grid, and the availability of electricity [12].

3.2.3.2 Environmental Issues

When discussing the environmental effects of EVs, carbon dioxide emissions are often cited as a representative of GHG emissions. In comparison to ICEs, EVs produce fewer GHG emissions and fewer emissions from the wheels themselves [13]. Not all situations are like that. Charging electric vehicles at peak demand, when most energy is generated by coal-based facilities, will require the burning of more coal to provide the necessary amount of additional electricity, which in turn will raise GHG emissions. Using the charger during non-peak times results in less pollution being produced. Further reductions are achieved by the utilization of renewable energy sources. There may be an increase in carbon dioxide (CO_2) emissions if more people use EVs charged by coal and other conventional energy sources. [14]. According to the findings of another study [15], as compared to ICEs, electric mobility systems employing an evening charge method with conventional energy result in a 10% reduction in CO_2 emissions.

3.2.3.3 Economic Issues

A DC rapid CS, while convenient, is substantially more expensive than a standard CS. In Table 3.2, we can see that a public level 2 CS will set you back over $3000 and a DC fast CS would set you back over $9000 [16].

3.3 LOCATION PROBLEM FORMATION APPROACHES

The author has analyzed the existing literature and concluded that positioning the CSs in such a way as to reduce the installation costs and increase earnings from charging EVs is essential for CS investors. EV drivers desire to reduce their travel costs, charging times, waiting times, charging costs, access fees, and so on, while distribution network operators aim to lessen the influence of CSs on the parameters

---

TABLE 3.2

Cost of chargers (Source: [16])

Charging Technologies	Charging Time	Cost of Installation
Level 1	4–11 hours	800–900$
Level 2	1–4 hours	1000–3000$
Level 3	≤30 minutes	70,000–75,000$ (for a 150 kW charger)

of the distribution system. A schematic of the overall framework for identifying optimal locations for CSs can be seen in Figure 3.3. Further, this work covers three strategies for optimal EVCS deployment [17], as depicted in Figure 3.4.

3.3.1 Distribution Network Operator (DNO) Approach

The distribution systems are in charge of delivering electricity to all connected electric loads in homes, businesses, and factories. The additional loads would influence the distribution network parameters. As a result, the DNO approach optimizes the

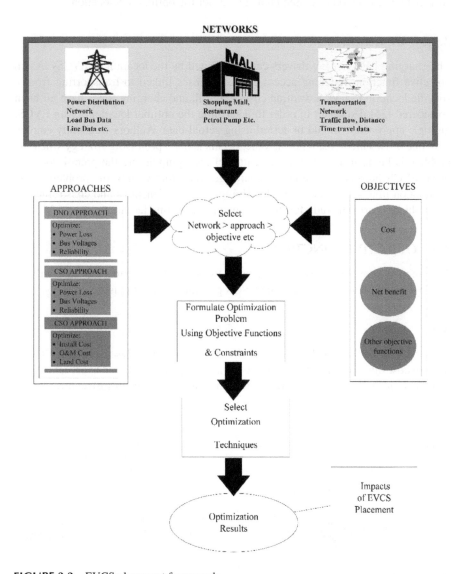

FIGURE 3.3 EVCS placement framework

distribution system's active power loss cost [18, 19], reactive power loss cost, voltage deviation cost [20], and costs linked with system reliability and stability [21].

3.3.2 CHARGING STATION OWNER (CSO) APPROACH

The CSO is liable for all expenses incurred during the installation of an EVCS in order to maximize revenue from EVs charging at the station. As a result, CSOs are looking for high-profit, low-overhead CS sites. Therefore, the CSO approach considers the investment cost [22], installation cost [23], operating cost [24], maintenance cost, road construction cost, and land cost [2] for the optimal CS location.

3.3.3 EV USER APPROACH

The charging habits of EV drivers can be affected by the location of EVCSs. As an objective function, reducing access costs, expenses linked to transporting from a specific location to EVCS, waiting costs [22, 24], and charging times have all been considered [25]. However, in order to determine the optimal location for an EVCS, multiple approaches should be tested using actual data. Authors frequently employ only one or two methods, ignoring others. This is not a promising strategy when the problem is being formulated. As a matter of fact, pinpointing the precise location of an EVCS presents a formidable challenge when formulating the problem of its installation in a designated area. Identifying objective functions and constraints for the problem is crucial study for the placement of charging facilities.

3.4 OBJECTIVE FUNCTION

This section provides a brief overview of the several goal functions that will be employed in the formalization of the EVCS placement planning issue.

3.4.1 COST

Numerous studies have included cost as a key factor in their analyses. Figure 3.4 shows that the cost functions may be formed using a variety of criteria and approaches. The construction of an EVCS requires a one-time investment in infrastructure, which

FIGURE 3.4 Techniques of problem formulation for EVCS placement

may be broken down into land and building costs as well as charger and labour expenses. The yearly cost of the power utilized to supply the charging services is also included in the operational cost. An objective function suggested in various research studies [23, 26] for the deployment of EVCSs is defined in Equation (3.1)

$$DC_i = C_{initial} + 25 \times C_{land} \times NC_i + PC \times C_{connector} \times (NC_i - 1) \tag{3.1}$$

Where, $C_{initial}$ is the initial cost of the charging facility, C_{land} is the cost of property, NC_i is the connection count at the ith EVCS, and $C_{connector}$ is the charging connector. When calculating the total cost of installing an EVCS, researchers discovered that transportation expenses between the charging demand point and the EVCS were the most significant. According to Equation (3.2), the objective function for the user cost of an EV is expressed as below [19, 26].

$$USEV = \sum_{x=1}^{FCS} \sum_{y=1}^{ZONE} dit(y,x) \times SPEC \times \sum_{z=1}^{24} PSEV(z) \times NEV(y) \times EP \tag{3.2}$$

Where FCS and ZONE are the number of fast-charging facilities and study area zones, respectively, $dit(y, x)$ is the distance between potential charging facility and zone locations, $PSEV(z)$ is the probabilistic set of EV charging in z hours, and SPEC is the particular amount of power that EVs use. To further determine the ideal location for a fast-charging facility, the authors [24, 26] formulated Equation (3.3) to calculate the yearly electric cost of the fast-charging facility.

$$FCSC_E = \sum_{s \in S} \sum_{i_F=1}^{I_F} \sum_{a \in J_{i_F}} \left(Days_s \times \frac{(SOC_a^d - SOC_a^g) \times B_a}{\eta} \times CE \right) \tag{3.3}$$

Where $Days_s$ is the total number of days, B_a is the capacity of the bank in the ath EV, SOC_a^d and SOC_a^g are the SOC at the start and finish of the trip, and CE is the price of energy per kWh.

3.4.2 NET BENEFIT

The net profit is employed as the objective function in the design of EVCSs with V2G connectivity (Figure 3.5). Financial savings can be realized in EVCSs by buying electricity directly from EV owners at a lower cost during peak power demand periods. By discharging their short-term energy storage, electric vehicles help the grid and V2G scheme. Electric vehicle charging at night is more cost-effective due to lower electricity rates and use. The EVCSs can increase their earnings by offering daytime charging at more competitive rate compared to overnight charging. Furthermore, one advantage of using power from the upstream grid is the revenue that EVCSs produce in Equation (3.5) by delivering electricity to the grid during times of high demand [26, 27].

$$Rev(i) = EMP_p \times P_{park}(i) \times t_{discharge}(i) \tag{3.4}$$

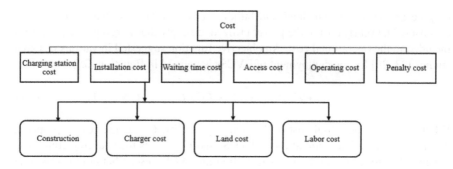

FIGURE 3.5   Different types of cost objective functions

Where EMP_p is the energy market price during peak hours, $t_{discharge}(i)$ is the cumulative time an EV cell is discharged via a V2G service, and $Rev(i)$ is the cumulative income received from the ith charging facility. Improved measured data and voltage profiles for EVCSs are one of the benefits of using the V2G method and may be translated into greater reliability and voltage profile income, as shown in Equation (3.5) [26, 28].

$$C_{ENS}(j) = \left[\sum_{b=1}^{N_{line}} E_{nsj} \times y_B \times L_B \times \left(\sum_{isf=1}^{N_{isf}} P_{isf} \times t_{isf} + \sum_{isfr=1}^{N_{isfr}} P_{isfr} \right) \times t_{isfr} \right] + C_{eqpf\,j} \qquad (3.5)$$

where N_{line} is the number of lines in the network, E_{nsj} is the rate of electricity not given to load at j, y_B is the probability of branch failure, L_B is branch length, N_{isf} is the total number of isolated endpoints during the outage, N_{isfr} is the number of secluded end nodes during maintenance, P_{isf} is the load not supplied during maintenance, t_{isf} is the fault repair time, P_{isfr} is the load not supplied for the duration of fault repair, t_{isfr} is the time for fault maintenance, and C_{eqpfj} is the cost of the energy not supplied because of equipment failure, with the exception at branch j. If a parking lot positioned in a distribution system might be used as a backup power source to bring electricity back for some of the failed loads, this would improve the distribution system. For each year distribution company, the benefit of greater reliability could be evaluated using Equation (3.6), as described in [26, 29].

$$DC_{NS}(j) = C_{ENS}(j) - C_{ENSV2G}(j) \qquad (3.6)$$

where $C_{ENS}(j)$ is the energy not supplied cost without V2G and $C_{ENSV2G}(j)$ is the energy not supplied cost with V2G.

3.4.3   OTHER OBJECTIVE FUNCTIONS

The researchers also take into consideration power loss, distance travelled, and the power supply moment balancing index while addressing placement planning issues for CSs, along with the previously specified objective functions. A higher loss of electricity would occur from an increase in load. Consequently, to minimize power loss, EVCSs must be carefully placed inside the distribution network. The key factors for placing EVCSs under the DNO approach include the price of power loss and the price of voltage deviation. As a result, most researchers have added power loss prices [24] and voltage deviation prices [28] as objective functions in Equations (3.7–3.8). For the power flow analysis, the Gauss–Seidel method, Newton–Raphson method, fast decoupled method, backward–forward sweep algorithm, and direct approach-based algorithm are used [29]. Due to its numerous benefits, the backward–forward sweep algorithm and its modified variant [26, 30] have been utilized for power flow analysis in numerous literary works.

$$P_{loss}^c = E_{cost} \sum_{i=1}^{NB} \sum_{j=1}^{NB} g_{ij} \left(V_i^2 - V_j^2 - 2V_i V_j \cos\left(\theta_{ij}\right) \right) \tag{3.7}$$

where E_{cost} represents the price of one unit of electricity in dollars, NB represents the line number, g_{ij} is the line conductance from the ith to the jth bus, V_i is the voltage at the ith bus, θ_{ij} is the load angle difference, VD^t is the voltage deviation at time t, and V_{buth}^t is the voltage at the b_{uth} bus.

$$VD^t = \sum_{t=1}^{24} \sum_{buth=1}^{N_{buth}} abs\left(1 - V_{buth}^t\right) \tag{3.8}$$

The power supply moment balancing index is a measurement that may be used to gauge the extent of power supply variation.

3.5 CONSTRAINTS

A set of equality and inequality constraints are applied to the CS position planning issue, as shown in Figure 3.6. After EVCSs are introduced to the distribution system, it is necessary to adhere to strict voltage, current, and temperature constraints at all buses. The installation range for EVCSs, both in terms of maximum and minimum values, must also be indicated. EVCSs should not be clustered together. The different types of equality and inequality constraints are as described below.

3.5.1 VOLTAGE CONSTRAINTS

A voltage limit for buses that accounts for inequality is stated in Equation (3.9), where V^{min} and V^{max} represent the voltage limitations at the Jth bus [23].

$$V_J^{min} < V_J < V_J^{max} \tag{3.9}$$

3.5.2 ACTIVE AND REACTIVE POWER CONSTRAINTS

The distribution system's energy requirements (both active and reactive) should be balanced [26, 31]. Therefore, in Equations (3.10) and (3.11), researchers formulate an

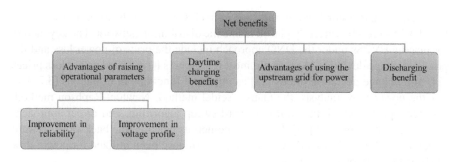

FIGURE 3.6   Value analysis of the use of V2G in the deployment of EVCS

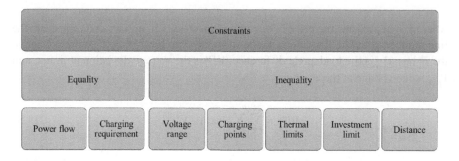

FIGURE 3.7   EVCS placement constraints for problem formulation

equality constraint on the power. P_{GI} and Q_{GI} represent the active and reactive power drawn from the grid, respectively; P_{DI} and Q_{DI} represent the active and reactive power demand of the distribution network, respectively; V_I and V_J are the voltages of Ith and J^{th} bus, respectively; and Y_{IJ} is the admittance of line from the Ith to the Jth bus.

$$P_{GI} - P_{DI} - V_I \sum_{J=1}^{n} V_I Y_{IJ} \cos\left(\gamma_I - \gamma_J - \theta_{IJ}\right) = 0 \tag{3.10}$$

$$Q_{GI} - Q_{DI} - V_I \sum_{J=1}^{n} V_I Y_{IJ} \cos\left(\gamma_I - \gamma_J - \theta_{IJ}\right) = 0 \tag{3.11}$$

3.5.2.1 Power Inequality Constraints

On the ith bus, a limit is placed on the real and reactive power's minimum and maximum values, as indicated by Equations (3.12) and (3.13) [23, 26].

$$P_{GI}^{min} < P_{GI} < P_{GI}^{max} \tag{3.12}$$

$$Q_{GI}^{min} < Q_{GI} < Q_{GI}^{max} \tag{3.13}$$

3.5.2.1 Branch Current Constraints

Each node in the distribution system must maintain the maximum allowable current [26], which is given by Equation (3.14).

$$I_{Br} < I_{Br}^{max} \qquad (3.14)$$

3.5.3 SOC OF THE BATTERY

When charging or discharging, it's important to keep the SOC [26] between its maximum and minimum values, as calculated by Equation (3.15), in order to maintain the battery health of an EV.

$$25\% \leq SOC \leq 90\% \qquad (3.15)$$

3.6 METHODS FOR OPTIMAL LOCATION PLANNING OF AN EVCS

Finding a charging facility may be done through a number of different methods, but they can all be categorized as either node-based, path-based, or tour-based [32]. The node-based approach has been shown to be the most efficient for identifying potential locations for CSs. It views the question of location as a singular situation of the more general CS site problem. This is the formalization of the issue at hand. Given possible locations, which are called "nodes," the goal is to find facilities, in this case CSs, at the nodes that meet the need. Appropriate approximations may be found using heuristic procedures in a reasonable period of time; hence, these are frequently employed. An explanation of the method's basic principle is provided in Figure 3.7. The path-based method, described in [33] and depicted in Figure 3.8, is a second strategy that has been considered. The idea behind this method is to ensure that the maximum number of people can be served by installing CSs in high-traffic areas and considering the direction in which vehicles are travelling. In contrast to the node-based approach, which gives a pretty static view of demand, this method takes into account only the effects that arise from demand caused by vehicle flows. Finally, as shown in Figure 3.9, the tour-based method considers an agent's and a vehicle's total activity over a time period, not just individual origin–destination trips. It considers

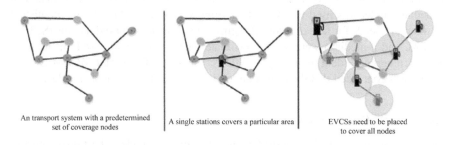

An transport system with a predetermined set of coverage nodes | A single stations covers a particular area | EVCSs need to be placed to cover all nodes

FIGURE 3.8 Working principle of the node-based approach

where users come from, how far they travel, where they go, and how long they stay in one place to determine where to place charging infrastructures.

3.6.1 NODE-BASED APPROACH

Facility location models like the set-covering location model (SCLM) aim to reduce the necessary number of facilities while still meeting all customers' needs [34]. The facilities in this model are spread out so that no demand nodes are farther than some maximum distance from a central power plant. It guarantees that everyone within a certain radius of a given point will have access to some kind of service, but it doesn't take demand into account; all demand points are treated equally, and it's enough that they're all covered. Using a mixed integer programming method based on vehicle routing logics, [35] proposed a model for the location of refueling stations to ensure that all nodes in a transportation network are reachable. When the capacity of each station is infinite, [36] used an SCLM to accommodate the highest demand for both local and long-distance travel at the lowest possible cost. Another approach that makes use of the network nodes is the maximum-covering location model (MCLM) [37]. Similar to the SCLM, this approach aims to optimize demand handling by strategically deploying a definite number of charging facilities so that no demand nodes are further than the defined critical distance from any of them. The MCLM is preferred over the SCLM because it allows for some demand nodes to not be covered when resources are insufficient to cover all of the demand nodes, as is often the case in practice. Although both the SCLM and the MCLM treat charger placement at a demand node or at the node's critical distance as equivalent, neither model takes into consideration the effect of distance when deciding whether the demand node is covered. Flow-capturing vs. node-based maximum coverage models for finding the optimum site of fast-charging stations are presented in [38, 39]. Maximum coverage optimization was used in [40] to estimate the monetary value of installing CSs in high-traffic areas like universities and shopping malls. Since its introduction in [41], the p-median model has become popular for solving CS location problems. For a p-median problem, the objective is to allocate each customer to a facility in such a method that the transportation costs (or the metric distance) between consumers and production facilities are minimized. Capacity constraints imposed by the available infrastructure mean that the problem can be solved; in this case, the demand from customers assigned to a given facility will never rise above that facility's ability to meet it. This means that the size of a CS determines the number of vehicles that can be charged in a given time frame. The p-median model was used to find fast-charging stations, and [42, 43] distinguished between the needs for fast and slow charging. The three node-based models (SCLM, MCLM, and p-median) were used to estimate charging demand based on Beijing's socio-demographic data as an input. After comparing SCLM and MCLM results, researchers concluded that the p-median model provides more reliable results [44].

3.6.2 PATH-BASED APPROACH

As an alternative to dealing with demand at individual nodes, the flow-capturing location model (FCLM) was presented [33]. This is a path-based variation of the MCLM based on the idea that traffic in a distribution network can be handled by numerous chargers placed on common pathways. The FCLM looks at sets of origins and destinations and tries to maximize the amount of traffic that is counted along the shortest route between them. If a route includes at least one node that is connected to a CS, then the entire route is considered covered by the scheme. For this reason, the FCLM was extended. Considering the limited range of alternative-fuel vehicles and the fact that a vehicle may need to stop at more than one refueling station in order to finish a trip, [45] devised the flow-refueling location model (FRLM) for alternative-fuel vehicles. They realized that it wouldn't be enough to just position CSs at nodes; therefore, they came up with a plan for station placement along linkages [46]. Then they developed the CFRLM with [45], which is a FRLM with capacity constraints. Previous models only considered one type of CS, so [47] used this model to strategically locate a variety of stations. Then, [48] created a model for performance optimization that accounts for detours from the shortest road that drivers should have to travel to refill their car, and [49] offered a model that allows for various deviation pathways. Taking into account the network's dynamic nature over time, [50] proposed a multi-path FRLM with the intention of controlling the cost of installation while ensuring that every trip can be completed via a route between the origin and the destination within an acceptable tolerance, especially in comparison to the shortest route. To further account for the unpredictability of the demand for EV charging after the requisite infrastructure is in place, [51] created a stochastic FCLM model.

3.6.3   Tour-Based Approach

The last option is the tour-based approach, which is also referred as the activity-based approach. Estimating the need for CSs for cars based on parking requirements is one of the central ideas behind the model proposed in [52]. The more slots are filled, the authors reasoned, the greater the charging demand will be, regardless of turnover. A parking-based model was created in [53] that takes into account total parking time but does not include parking at one's own residence [54, 55]. The number of journeys that could not be completed due to absence of a charging facility is reduced using their optimization model. The authors of [56] considered the current status of charging technologies while formulating their method for missing trips. Using the existing groundwork, they developed a "user charging model" to determine where and how EV owners should charge. An optimization software provided data about where EVs are unable to finish trips owing to a shortage of charging facilities. Using data from taxi rides, [57] proposed a technique for incorporating CSs into the existing gas station network. By analyzing the current level of charge in relation to prior tours and extracting stop events, researchers have evaluated the prospective charging demand for stop sites in petrol stations [58–60]. Further, cannibalization occurs when CSs overlap each other's coverage areas, which is an issue that arises from traffic flows but is ignored by a node-based approach. Moreover, [61] discovered that the flow-based strategy is more reliable when the number of fast chargers to locate rises, which is crucial for long-term planning. That's why many studies factor in traffic patterns when deciding where to put CSs. Even though this flow-based strategy has its uses, it cannot be applied to everything. Flow-based approaches assume that, like users of ICE vehicles, EV drivers will quickly charge their vehicles and then continue

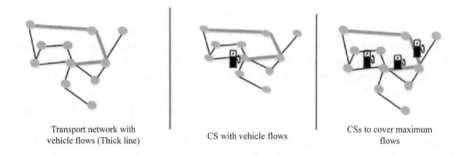

Transport network with
vehicle flows (Thick line)   CS with vehicle flows   CSs to cover maximum
                                                                                       flows

FIGURE 3.9 Working principle of the path-based approach

to their destination. Fast-charging stations, which can fully replenish the battery of an electric vehicle in about 12 minutes, are necessary for this system to work. Therefore, the flow-based approach is not a substitute for the node-based strategy but rather a supplement to it, depending on the objective, region, CS type, etc. However, academics rarely use both terms together. In light of the drawbacks of the flow-based method, the tour-based method is predicated not just on users' driving habits but on their actions in general. The term "activity-based" is sometimes used to describe this method as well. It provides a more accurate representation of users' charging requirements than node- or path-based methods by taking into account events surrounding the specifics of the sequence of trips. Tour-based approaches, which employ real and individual data, capture the randomization of user behavior and provide service to all users, which is difficult with aggregated data, as shown in Figure 3.9. Most tour-based works, however, make it clear that this approach is typically data-driven, making use of either real or simulated data. To create a realistic model, a massive amount of detailed data is needed, down to the level of individual trips and stops at the very least. Often, it is challenging to acquire such information.

3.7 SOLUTION ALGORITHMS FOR OPTIMAL LOCATION PROBLEMS

By formulating an optimization problem, we can use optimization methods to either minimize or maximize the associated objective function. As depicted in Figure 3.10, the cost function is highly amenable to optimization using a variety of available optimization methods. A wide variety of optimization problems, both linear and nonlinear, convex and concave, may be formulated for the placement of EVCS. Based on the used variables, the framed problem can be continuous, integer, discrete, and combination of these types. This means that precise choice of optimization method for a specific problem is of utmost importance [26].

In addition, there are two primary types of optimization strategies: classical and advanced. Using conventional optimization techniques, we can discover the global highest and lowest values of a function that is continuous and differentiable, with no further constraints. The traditional approaches' reliance on decision variables that are

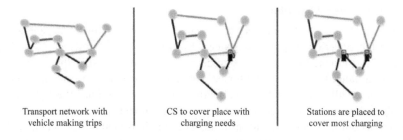

Transport network with vehicle making trips

CS to cover place with charging needs

Stations are placed to cover most charging

FIGURE 3.10 Working principle of the tour-based approach

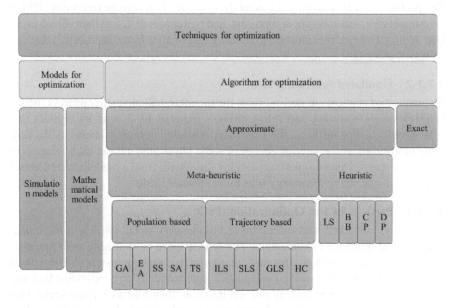

FIGURE 3.11 Classification of optimization techniques

not continuous and/or differentiable also limits their practical use in some contexts. The optimization of large-scale problems is intrinsically linked to the multi-modality, high-dimensionality, and differentiability of modern optimization methods, all of which are intractable to traditional approaches. Because most classical methods rely on gradient information, they cannot be used to solve functions that are not differentiable. Furthermore, optimization problems with multiple local optima are notoriously difficult for classical methods to resolve. Modern optimization methods, however, get around these obstacles to find a solution to the optimization issue.

3.7.1 OPTIMIZATION TECHNIQUES FOR SINGLE-OBJECTIVE FUNCTIONS

An optimization problem with a single objective is much simpler to resolve than one with several objectives. Consequently, single-objective optimization problems are resolved employing both traditional and cutting-edge optimization algorithms.

3.7.1.1   Genetic Algorithm

Genetic algorithms (GAs) model the parental traits of candidate populations to improve the selection of previous groups. In order to successfully apply a GA, it is necessary to make thoughtful design decisions in order to tailor the algorithm to the issue at hand. In actuality, the fitness functions, crossover process, and gene-encoding system all contribute to how effectively the algorithm finds the right answer. The algorithm must have access to a significant amount of diverse data in order to avoid becoming stuck in local minima. The standard method for accomplishing this is to randomly select genes for a crossover, which slows the convergence rate but guarantees exploration. Even a modest improvement in the GA solution caused by increasing the population size results in a huge increase in computation time. The problem in [62] is formulated using the objective functions of travel cost, EVCS investment cost, substation operation cost, and power loss cost, and then solved using GA. A mixed integer nonlinear programming (MINLP) is formulated in [63] that can be solved using GA.

3.7.1.2   Simulated Annealing (SA)

The name and concept are derived from the metallurgical annealing process, which involves slowly warming and refrigerating a material to enlarge its crystals and remove defects. For instance, the simulated annealing approach interprets the task to be minimized as the interior energy of some physical system at each location in the search space. This requires transitioning the system from a predetermined initial condition to one where the energy is as low as possible [64].

3.7.1.3   Particle Swarm Optimization (PSO)

PSO, a new, well-liked, and successful technique, uses global particle communication and real-number randomization. The search space is swarming with potential solutions (particles), each sharing and comparing its own best answer with the finest solutions from across the world. Each particle determines its own individual best and the overall best at the beginning of each iteration, then follows a route vector that converges with the overall best. A more recent and accurate PSO variation called IPSO was created to shorten computation times. The PSO technique is utilized to determine the ideal position for the radial distribution system (RDS) to install the EVCS and distributed energy resources (DERs) because power loss is the optimization problem's objective function in [65].

3.7.1.4   Teaching–Learning-Based Optimization Algorithm

The teaching–learning-based optimization (TLBO) algorithm, which aims to maximize student performance in a classroom context, is inspired by the process of teaching and learning. The algorithm also identifies two main modes of instruction: (i) direct instruction from a teacher (the "teaching cycle") and (ii) peer-to-peer contact with other students (known as the learning phase). A hybrid CSO/TLBO algorithm is used to discover the best site for EVCS. The algorithm aims to minimize the costs associated with the EVCS, voltage irregularities, system reliability, and the EVCS accessibility index [66].

3.7.1.5 Gray Wolf Optimization

Optimizing like a grey wolf was conceptualized by [66]. The author draws inspiration from the grey wolf's natural behavior and hunting methods. In addition, grey wolves have a unique pack hierarchy. Each pack has a dominant member known as the alpha wolf. When it comes to the next grouping, grey wolves come in at a respectable second place. They simplify matters for the alphas. They are called "beta wolves" for short. Wolves with a lower pack rank, or "delta wolves," are not as highly prized as those with higher ranks. Their objective is to dominate the omegas while gaining the respect of the alpha and beta packs. The omega grey wolves are the least significant pack because they are the alpha wolf's underlings.

3.7.1.6 Artificial Bee Colony (ABC) Algorithm

The ABC algorithm was developed by mimicking how real bees behave when searching for food sources like nectar and communicating their discoveries with more bees in the colony. The ABC is composed mostly of three dissimilar kinds of bees: workers, observers, and scouts. Each of them participates in the process differently by hovering from one place to another in a multi-dimensional search space that simulates the elucidation space. By minimizing power loss costs, power from grid costs, power from DER costs, and garage charging/discharging costs, the ABC algorithm and firefly algorithm (FA) in [67] are utilized to determine the best car parks lots in the distribution network.

3.8 OPTIMIZATION TECHNIQUES FOR MULTI-OBJECTIVE FUNCTIONS

Both an a priori and an a posteriori approach can be used for multi-objective optimization. In an effort to simplify a complex multi-objective optimization issue, a priori approaches combine goals. Also, a problem domain expert provides a set of weights that denote the relative importance of the objectives and the frequency with which they are provided. The fact that such approaches frequently require several algorithm iterations to find the Pareto optimal set is a significant disadvantage. Additionally, it takes a lot of work and is impossible to locate every Pareto optimal front on your own. The capability to determine the Pareto optimum set in a single run is the basis for the posterior approaches while maintaining multi-objective formulations of multi-objective issues. One more perk is that these techniques can be used to find any Pareto front. The trade-off is a higher computing expense and the need to pursue multiple goals at once [68].

3.8.1 NON-DOMINATED SORTING GENETIC ALGORITHM II

Meta-heuristic multi-objective genetic algorithms like non-dominated sorting genetic algorithm II (NSGA-II) are widely utilized to tackle multi-objective optimization problems in contexts including facility distribution, supply network design, and crowded facility placement. On the basis of low dominance, the NSGA-II divides the population into several "chromosome-front groupings."

Furthermore, NSGA-II is used to formulate and solve the multi-objective function that takes into account the costs of EVCS development, EV-specific energy consumption, electrical network power loss, DER power generation, and the maximum voltage deviation for EVCS and DER placement in the distribution network.

3.8.2 Multi-Objective Colliding Optimization Algorithm

It is the dynamics of collisions between physical things that provide inspiration for the evolutionary strategy recognized as colliding bodies optimization (CBO) [69]. Positive findings have been found by CBO for both restricted and unrestricted test functions, as well as for engineered single-objective problems. The lack of memory usage and the absence of fine-tuning parameters make this formulation of the algorithm easy to implement. In order to maximize profits while minimizing costs, researchers have developed a non-dominated, sorting-based, multi-objective CBO [70] method for optimizing the costs of carbon dioxide emissions and reinforced concrete structural components.

3.8.3 Multi-Objective Ant Lion Optimizer

The authors of [68] suggest a novel meta-heuristic technique dubbed multi-objective ant lion optimization (MOALO) to solve this problem. It works much like the natural relationship between ants and antlions. Antlions, which belong to the family Myrmeleontidae, are voracious insect predators that focus on ants as a primary food source. Moreover, antlions wait for their favored food in sand holes they have dug before snatching it up with their huge jaws. Understanding the modeling of single-objective ALO, as proposed in [70], is essential to grasping the full mathematical modeling of the MOALO algorithm.

3.9 EV LOAD INTEGRATION IMPACT ANALYSIS

The possible groups of EV integration impacts are showed in Figure 3.10. The initial set of consequences is the impact of EV load on distribution network characteristics, tailed by the effects on the economy and the environment.

3.9.1 EV Load Integration Impact on Distribution Networks

As EV charging infrastructure grows, many issues arise in the current distribution system. Recent years have seen thorough evaluations of these concerns. Current definitions of EV impact studies primarily concentrate on the impacts of electric cars on the dependability of energy production, the lifespan of transformers, and the power efficiency of distribution networks. Peak demand may rise if electric cars are charged during peak hours, calling for more power plants to be online. The increasing EV load demand would also affect the lifetime of the substation and service transformers. In addition, EV charging may cause voltage dips, power imbalances, and voltage/current harmonics, all of which degrade power quality. The possible impact of EV load integration could be categorized as shown in Figure 3.12.

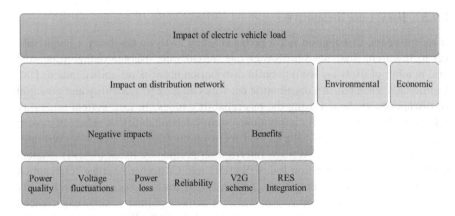

FIGURE 3.12 Impact of electric vehicle's load

3.9.2 NEGATIVE IMPACTS

Electric vehicle chargers are frequents culprits when it comes to grid compatibility issues. The IEEE 519 standard specifies that a total harmonics distortion (THD) value of less than 5% be met in order to guarantee uninterrupted operation for power networks up to 69 kilovolts (kV). There is evidence from this study [71, 72] to imply that harmonic disturbances would rise if EVCSs were integrated into the power grid. The total harmonic distortion is around 4.82% with a single electric vehicle (EV) hooked into the device, 12.35% with three EVs, and 19.69% with five EVs [71, 72].

3.9.3 IMPACT ON THE VOLTAGE

In this subdivision, we'll look at how integrating plug-in electric vehicles (PEVs) into the grid can affect voltage, and thus the efficiency with which electricity is transmitted to homes. The voltage drop at buses caused by charging is directly proportional to the load of EVs added to the distribution network. According to literature, there are areas where the voltage is down to less than 96% of the nominal value. This necessitates the implementation of system upgrades. Voltage deviation from rated voltage ranges from 12.7% to 43.3% at different charging rates between 20% and 80% PEV penetration, as shown in [32].

3.9.4 IMPACT ON POWER LOSS

When planning for upcoming charging requirements caused by the slow but continuous addition of PEVs into the grid, power system losses become a key challenge. Energy losses may rise by as much as 40% during off-peak charging, according to [73], assuming a 62% market penetration of PEVs. According to the researchers, the more the penetration of PEVs, the greater the network power losses. The power loss increment may possibly be reduced to a finite level by adopting the EVCS method in the most advantageous area.

3.9.5 Impact on Reliability

In recent years, distribution network consistency analysis has been a tough subject to research. Statistics on disaster rate, repair rate, average outage time, and total number of users are used to build distribution network reliability indices [18]. Additionally, each bus in a distribution network can have its reliability and potential for disruptions evaluated using the bus reliability index. In fact, system reliability indices evaluate the overall distribution system's dependability. A subset of system reliability indices includes metrics that focus on customers and energy consumption, respectively. System Average Interruption Frequency Index (SAIFI), System Average Interruption Duration Index (SAIDI), and Customer Average Interruption Duration Index (CAIDI) are all elegant acronyms for customer-centric reliability indices. The SAIFI metric measures the total number of interruptions experienced by system customers over a given time period, while the SAIDI metric specifies the mean duration of interruptions experienced by each customer. It is vital to note that both the length of the outage and the overall number of consumers have an impact on SAIDI [31].

3.10 POSITIVE IMPACTS

3.10.1 Benefit of V2G Schemes

Cheaper EV user prices, lower EVCS operator expenses, and flatter EVCS load curves are just a few of the benefits of V2G EVCS deployment. The capacity to deliver extra energy from a vehicle's battery to the power system during the highest demand periods and subsequently replenish the battery during the off hours is the most essential feature of the V2G programme. In addition, the sales demand for the form of electricity V2G power generates is what ultimately determines the revenue generated. In electricity-only markets, such as peak power, income is calculated by adding the price per kilowatt-hour to the quantity of energy actually sent. Additionally, the ability of V2G capacity would provide a fraction of the peak power, decreasing the necessity for the grid to procure electricity from wholesale market. Therefore, in [74], it was possible to calculate savings from using V2G power to supply loads instead of buying power on the wholesale market.

3.10.2 Easy Handling of Renewable Generation

Because of the intermittent nature of non-conventional energy, power companies are experiencing difficulty absorbing substantial volumes of renewable energy supply into their systems. The rapid-response control electronic interface of the EV charger, in conjunction with the energy storage provided by the batteries, shows that it is possible to deal with the intermittent nature of the power supply. Most encouragingly, the study found that 59% of the grid's generating capacity could be supplied by wind energy when EVs were used for primary frequency control. EVs, however, can be charged with a limited amount of solar power. Furthermore, the electricity needs of a sport utility vehicle are calculated using the typical 40-mile daily commute across North America.

3.10.3 IMPACT ON THE ENVIRONMENT

Because EVCSs use a distribution network to supply the power needs of EVs rather than fuel based on outdated technologies, carbon emissions are decreased. Pollutant emissions can be further reduced by charging a world run by EVs with green energy systems. When the whole lifespan of an electric vehicle, including manufacture and transportation, is included, the average yearly CO_2 equivalent emissions for an EV are 4,450 pounds [19]. Traditional diesel engines, on the other hand, may emit more than twice as much pollution each year. Furthermore, the location and most common energy sources for power are the key factors of the well-to-wheel pollution produced by your EVs.

3.10.4 ECONOMIC IMPACT

EVs may have a financial impact on both the EV owner and the electricity company. In particular, EVs are expensive in comparison to ICE cars. Conversely, EVs have lower fuel consumption and operational costs than ICE vehicles due to superior efficiency of electric motors. Further, the efficiency of ICE vehicles is typically between 15% and 18%, while that of EVs is typically between 60% and 70%. If the V2G concept can successfully transfer the electricity from an EV's storage bank to the grid, the owner of the EV will benefit. Basic power system costs for EV fleets are predicted to decrease by USD $200–300 per vehicle, per year, according to studies [74].

3.11 SUMMARY

A number of optimization methods were found to be useful in determining where EVCSs should be placed. This chapter compares and contrasts the efficacy of several optimization approaches for resolving EVCS placement challenges. Researchers have also examined a variety of factors to determine the optimal site for EVCS installation. Choosing the right objective functions, constraints, and solutions is crucial to the success of such techniques. There are a few different ways to go about charging electric vehicles: the DNO way, the CSO way, the EV user way, or a hybrid approach. The analysis conducted in [26] found that 15.2% of the research emphasizes the deployment of the DNO approach for positioning, while 6.52% of the study concentrates on the CSO approach for EVCS placement. Earlier studies primarily centered around the EV user approach, accounting for 4.35%, whereas the combination of DNO with CSO accounted for 32.6%. The combination of CSO with EV user comprised 15.22%, while EV user with DNO constituted 2.17%. The combination of all three approaches represented 26.1% of the problem formulation for EVCS placement, as depicted in Figure 3.13.

The most common methods used by researchers to solve EVCS placement problems are GA and PSO. Additional approaches to the location problem include ACO,

GA	Hybrid	PSO	TLBO	LP	BB	Others
22%	21%	18%	6%	3%	3%	27%

FIGURE 3.13 Percentage share of the optimization algorithms used in EVCS placement

ABC, TLBO, LP, greedy algorithm, GWO, GOA, branch and bound, and in-depth study approached. Additionally, we analyze how distribution system dependability is affected by voltage, power quality, power loss, and distribution system consistency. These studies focus on topics including distributed generation (DG) integration, uncertainty, and V2G schemes.

3.12 FUTURE SCOPE

3.12.1 RENEWABLE ENERGY INTEGRATED WITH EVCSS

Due to the intermittent nature of renewable energy production, electric utilities are experiencing trouble incorporating large volumes of renewable power sources into their systems. The highly adaptable electronic interface of the EV charger, combined with energy storage, can effectively handle the intermittent nature of renewable resources.

3.12.2 ACCURATE SOLUTION TECHNIQUES FOR EVCS PLACEMENT

The majority of studies have used GA and PSO to address the problem of EVCS location. Several approaches, such as grey wolf optimization, TLBO, artificial intelligence, and machine learning, might be included to provide a good and accurate alternative to the EVCS problem.

REFERENCES

1. Cook, J., D. Nuccitelli, S. A. Green, M. Richardson, B. Winkler, R. Painting, and A. Skuce. "Quantifying the consensus on anthropogenic global warming in the scientific literature." *Environmental Research Letters*, 8, no. 2 (2013), 024024.
2. Parker, Nathan, Hanna L. Breetz, Deborah Salon, Matthew Wigginton Conway, Jeffrey Williams, and Maxx Patterson. "Who saves money buying electric vehicles? Heterogeneity in total cost of ownership." *Transportation Research Part D: Transport and Environment*, 96 (2021), 102893.
3. Berkeley, N., Jarvis, D., and Jones, A. "Analysing the take up of battery electric vehicles: An investigation of barriers amongst drivers in the UK." *Transportation Research Part D: Transport and Environment*, 63 (2018), 466–481.
4. Domke, G. M., B. F. Walters, D. J. Nowak, J. Smith, S. M. Ogle, and Coulston, J. W. "Greenhouse gas emissions and removals from forest land and urban trees in the United States, 1990–2017." *Resource Update FS-178. Newtown Square, PA: US Department of Agriculture, Forest Service, Northern Research Station*, 4 (2019), 1–4.
5. Richardson, D. B. "Electric vehicles and the electric grid: A review of modeling approaches, impacts, and renewable energy integration." *Renewable and Sustainable Energy Reviews*, 19 (2013), 247–254.

6. Patnaik, L., A. V. J. S. Praneeth, and S. S. Williamson. A closed-loop constant-temperature constant-voltage charging technique to reduce charge time of lithium-ion batteries. *IEEE Transactions on Industrial Electronics*, 66, no. 2 (2018), 1059–1067.
7. Hemavathi, S. and A. Shinisha. "A study on trends and developments in electric vehicle charging technologies." *Journal of Energy Storage*, 52 (2022), 105013.
8. Medora, N. K. *Electric and Plug-in Hybrid Electric Vehicles and Smart Grids: In the Power Grid* (pp. 197–231). Academic Press (2017).
9. Shepero, M., J. Munkhammar, J. Widén, J. D. Bishop, and T. Boström. "Modeling of photovoltaic power generation and electric vehicles charging on city-scale: A review." *Renewable and Sustainable Energy Reviews*, 89 (2018), 61–71.
10. Raff, R., V. Golub, D. Pelin, and D. Topić. "Overview of charging modes and connectors for the electric vehicles." In *2019 7th International Youth Conference on Energy (IYCE)* (pp. 1–6). IEEE (2019, July).
11. Url: https://theicct.org/sites/default/files/publications/icct_ev_charging_cost_20190813.Pdf.
12. Gomes, I. S. F., Y. Perez, and E. Suomalainen. "Coupling small batteries and PV generation: A review." *Renewable and Sustainable Energy Reviews*, 126 (2020), 109835.
13. Thompson, A. W., and Y. Perez. "Vehicle-to-everything (v2x) energy services, value streams, and regulatory policy implications." *Energy Policy*, 137 (2020), 111136.
14. Akhavan-Rezai, E., M. F. Shaaban, E. F. El-Saadany, and A. Zidan. "Uncoordinated charging impacts of electric vehicles on electric distribution grids: Normal and fast charging comparison." In *2012 IEEE Power and Energy Society General Meeting* (pp. 1–7). IEEE (2012, July).
15. Ma, H., F. Balthasar, N. Tait, X. Riera-Palou, and A. Harrison. "A new comparison between the life cycle greenhouse gas emissions of battery electric vehicles and internal combustion vehicles." *Energy Policy*, 44 (2012), 160–173.
16. Nicholas, M. "Estimating electric vehicle charging infrastructure costs across major us metropolitan areas." *International Council on Clean Transportation*, 14, no. 11 (2019).
17. Hadley, S. W. "Impact of plug-in hybrid vehicles on the electric grid." *Ornl Report*, 640 (2006).
18. Gampa, Srinivasa Rao, Kiran Jasthi, Preetham Goli, D. DAS, and R. C. Bansal. "Grasshopper optimization algorithm based two stage fuzzy multiobjective approach for optimum sizing and placement of distributed generations, shunt capacitors and electric vehicle charging stations." *Journal of Energy Storage*, 27 (2020), 101117.
19. Chen, Liang, Chunxiang Xu, Heqing Song, and Kittisak Jermsittiparsert. "Optimal sizing and sitting of EVCS in the distribution system using metaheuristics: A case study." *Energy Reports*, 7 (2021), 208–217.
20. Awasthi, Abhishek, Karthikeyan Venkitusamy, Sanjeevikumar Padmanaban, Rajasekar Selvamuthukumaran, Frede Blaabjerg, and Asheesh K. Singh. "Optimal planning of electric vehicle charging station at the distribution system using hybrid optimization algorithm." *Energy*, 133 (2017), 70–78.
21. Amini, M. Hadi, Mohsen Parsa Moghaddam, and Orkun Karabasoglu. "Simultaneous allocation of electric vehicles' parking lots and distributed renewable resources in smart power distribution networks." *Sustainable Cities and Society*, 28 (2017), 332–342.
22. Simorgh, Hamid, Hasan Doagou-Mojarrad, Hadi Razmi, and Gevork B. Gharehpetian. "Cost-based optimal siting and sizing of electric vehicle charging stations considering demand response programmes." *IET Generation, Transmission & Distribution*, 12, no. 8 (2018), 1712–1720.
23. Battapothula, Gurappa, Chandrasekhar Yammani, and Sydulu Maheswarapu. "Multi-objective simultaneous optimal planning of electrical vehicle fast charging stations and DGS in distribution system." *Journal of Modern Power Systems and Clean Energy*, 7, no. 4 (2019), 923–934.
24. Kong, Weiwei, Yugong Luo, Guixuan Feng, Keqiang Li, and Huei Peng. "Optimal location planning method of fast charging station for electric vehicles considering operators, drivers, vehicles, traffic flow and power grid." *Energy*, 186 (2019), 115826.

25. Zhang, Hongcai, Zechun Hu, Zhiwei Xu, and Yonghua Song. "An integrated planning framework for different types of PEV charging facilities in urban area." *IEEE Transactions on Smart Grid*, 7, no. 5 (2015), 2273–2284.
26. Ahmad, Fareed, Atif Iqbal, Imtiaz Ashraf, and Mousa Marzband. "Optimal location of electric vehicle charging station and its impact on distribution network: A review." *Energy Reports*, 8 (2022), 2314–2333.
27. Moradi, Mohammad H., Mohammad Abedini, S. M. Reza Tousi, and S. Mahdi Hosseinian. "Optimal siting and sizing of renewable energy sources and charging stations simultaneously based on differential evolution algorithm." *International Journal of Electrical Power & Energy Systems*, 73 (2015), 1015–1024.
28. Rupa, J. A. Michline and S. Ganesh. "Power flow analysis for radial distribution system using backward/forward sweep method." *International Journal of Electrical, Computer, Electronics and Communication Engineering*, 8, no. 10 (2014), 1540–1544.
29. Sereeter, Baljinnyam, Kees Vuik, and Cees Witteveen. "Newton power flow methods for unbalanced three-phase distribution networks." *Energies*, 10, no. 10 (2017), 1658.
30. Petridis, Stefanos, Orestis Blanas, Dimitrios Rakopoulos, Fotis Stergiopoulos, Nikos Nikolopoulos, and Spyros Voutetakis. "An efficient backward/forward sweep algorithm for power flow analysis through a novel tree-like structure for unbalanced distribution networks." *Energies*, 14, no. 4 (2021), 897.
31. Deb, Sanchari, Xiao-Zhi Gao, Kari Tammi, Karuna Kalita, and Pinakeswar Mahanta. "A novel chicken swarm and teaching learning based algorithm for electric vehicle charging station placement problem." *Energy*, 220 (2021), 119645.
32. Deb, Sanchari, Kari Tammi, Karuna Kalita, and Pinakeswar Mahanta. "Review of recent trends in charging infrastructure planning for electric vehicles." *Wiley Interdisciplinary Reviews: Energy and Environment*, 7, no. 6 (2018), e306.
33. Hodgson, M. John. "A flow-capturing location-allocation model." *Geographical Analysis*, 22, no. 3 (1990), 270–279.
34. Toregas, Constantine, Ralph Swain, Charles Revelle, and Lawrence Bergman. "The location of emergency service facilities." *Operations Research*, 19, no. 6 (1971), 1363–1373.
35. Wang, Ying-Wei and Chuah-Chih Lin. "Locating road-vehicle refueling stations." *Transportation Research Part E: Logistics and Transportation Review*, 45, no. 5 (2009), 821–829.
36. Wang, Ying-Wei and Chuan-Ren Wang. "Locating passenger vehicle refueling stations." *Transportation Research Part E: Logistics and Transportation Review*, 46, no. 5 (2010), 791–801.
37. Church, Richard and Charles Revelle. "The maximal covering location problem." *In Papers of the Regional Science Association*, 32, no. 1 (1974), 101–118. Springer-Verlag.
38. Frade, Inês, Anabela Ribeiro, Gonçalo Gonçalves, and António Pais Antunes. "Optimal location of charging stations for electric vehicles in a neighborhood in Lisbon, Portugal." *Transportation Research Record*, 2252, no. 1 (2011), 91–98.
39. Sun, Zhuo, Wei Gao, Bin Li, and Longlong Wang. "Locating charging stations for electric vehicles." *Transport Policy*, 98 (2020), 48–54.
40. Wagner, Sebastian, Markus Götzinger, and Dirk Neumann. "Optimal location of charging stations in smart cities: A points of interest-based approach." (2013).
41. Hakimi, S. Louis. "Optimum locations of switching centers and the absolute centers and medians of a graph." *Operations Research*, 12, no. 3 (1964), 450–459.
42. Gavranović, Haris, Alper Barut, Gürdal Ertek, Orkun Berk Yüzbaşioğlu, Osman Pekpostalci, and Önder Tombuş. "Optimizing the electric charge station network of EŞARJ." *Procedia Computer Science*, 31 (2014), 15–21.
43. He, Sylvia Y., Yong-Hong Kuo, and Dan Wu. "Incorporating institutional and spatial factors in the selection of the optimal locations of public electric vehicle charging facilities: A case study of Beijing, China." *Transportation Research Part C: Emerging Technologies*, 67 (2016), 131–148.

44. An, Yu, Bo Zeng, Yu Zhang, and Long Zhao. "Reliable p-median facility location problem: Two-stage robust models and algorithms." *Transportation Research Part B: Methodological* 64 (2014), 54–72.
45. Upchurch, Christopher, Michael Kuby, and Seow Lim. "A model for location of capacitated alternative-fuel stations." *Geographical Analysis*, 41, no. 1 (2009), 85–106.
46. Kuby, Michael and Seow Lim. "Location of alternative-fuel stations using the flow-refueling location model and dispersion of candidate sites on arcs." *Networks and Spatial Economics*, 7, no. 2 (2007), 129–152.
47. Wang, Ying-Wei, and Chuah-Chih Lin. "Locating multiple types of recharging stations for battery-powered electric vehicle transport." *Transportation Research Part E: Logistics and Transportation Review*, 58 (2013), 76–87.
48. Kim, Jong-Geun and Michael Kuby. "The deviation-flow refueling location model for optimizing a network of refueling stations." *International Journal of Hydrogen Energy*, 37, no. 6 (2012), 5406–5420.
49. Huang, Yongxi, Shengyin Li, and Zhen Sean Qian. "Optimal deployment of alternative fueling stations on transportation networks considering deviation paths." *Networks and Spatial Economics*, 15, no. 1 (2015), 183–204.
50. Li, Shengyin, Yongxi Huang, and Scott J. Mason. "A multi-period optimization model for the deployment of public electric vehicle charging stations on network." *Transportation Research Part C: Emerging Technologies*, 65 (2016), 128–143.
51. Wu, Fei and Ramteen Sioshansi. "A stochastic flow-capturing model to optimize the location of fast-charging stations with uncertain electric vehicle flows." *Transportation Research Part D: Transport and Environment*, 53 (2017), 354–376.
52. Jia, Long, Zechun Hu, Yonghua Song, and Zhuowei Luo. "Optimal siting and sizing of electric vehicle charging stations." In *2012 IEEE International Electric Vehicle Conference* (pp. 1–6). IEEE (2012).
53. Chen, T. Donna, Kara M. Kockelman, and Moby Khan. "Locating electric vehicle charging stations: Parking-based assignment method for Seattle, Washington." *Transportation Research Record*, 2385, no. 1 (2013), 28–36.
54. Cavadas, Joana, Gonçalo Homem de Almeida Correia, and Joao Gouveia. "A MIP model for locating slow-charging stations for electric vehicles in urban areas accounting for driver tours." *Transportation Research Part E: Logistics and Transportation Review*, 75 (2015), 188–201.
55. You, Peng-Sheng, and Yi-Chih Hsieh. "A hybrid heuristic approach to the problem of the location of vehicle charging stations." *Computers & Industrial Engineering*, 70 (2014), 195–204.
56. Dogru, Mustafa, Matthew Andrews, John Hobby, Yue Jin, and Gabriel Tucci. "Modeling and optimization for electric vehicle charging infrastructure." In *24th Annual Pom Conference Denver*, 2013.
57. Cai, Hua, Xiaoping Jia, Anthony S. F. Chiu, Xiaojun Hu, and Ming Xu. "Siting public electric vehicle charging stations in Beijing using big-data informed travel patterns of the taxi fleet." *Transportation Research Part D: Transport and Environment*, 33 (2014), 39–46.
58. Shahraki, Narges, Hua Cai, Metin Turkay, and Ming Xu. "Optimal locations of electric public charging stations using real world vehicle travel patterns." *Transportation Research Part D: Transport and Environment*, 41 (2015), 165–176.
59. Gonzalez, Jairo, Roberto Alvaro, Carlos Gamallo, Manuel Fuentes, Jesús Fraile-Ardanuy, Luk Knapen, and Davy Janssens. "Determining electric vehicle charging point locations considering drivers' daily activities." *Procedia Computer Science*, 32 (2014), 647–654.
60. He, Fang, Yafeng Yin, and Jing Zhou. "Deploying public charging stations for electric vehicles on urban road networks." *Transportation Research Part C: Emerging Technologies*, 60 (2015), 227–240.

61. Wu, Fei and Ramteen Sioshansi. "A stochastic flow-capturing model to optimize the location of fast-charging stations with uncertain electric vehicle flows." *Transportation Research Part D: Transport and Environment*, 53 (2017), 354–376.
62. Xiang, Yue, Junyong Liu, Ran Li, Furong Li, Chenghong Gu, and Shuoya Tang. "Economic planning of electric vehicle charging stations considering traffic constraints and load profile templates." *Applied Energy*, 178 (2016), 647–659.
63. Sadeghi-Barzani, Payam, Abbas Rajabi-Ghahnavieh, and Hosein Kazemi-Karegar. "Optimal fast charging station placing and sizing." *Applied Energy*, 125 (2014), 289–299.
64. Eren, Yavuz, Ibrahim B. Küçükdemiral, and Ilker Üstoğlu. "Introduction to optimization." In *Optimization in Renewable Energy Systems* (pp. 27–74). Butterworth-Heinemann (2017).
65. Reddy, Moupuri Satish Kumar and Kamakshy Selvajyothi. "Optimal placement of electric vehicle charging station for unbalanced radial distribution systems." *Energy Sources, Part A: Recovery, Utilization, and Environmental Effects*, (2020), 1–15.
66. Mirjalili, Seyedali, Seyed Mohammad Mirjalili, and Andrew Lewis. "Grey wolf optimizer." *Advances in Engineering Software*, 69 (2014), 46–61.
67. El-Zonkoly, Amany and Leandro Dos Santos Coelho. "Optimal allocation, sizing of PHEV parking lots in distribution system." *International Journal of Electrical Power & Energy Systems*, 67 (2015), 472–477.
68. Mirjalili, Seyedali, Pradeep Jangir, and Shahrzad Saremi. "Multi-objective ant lion optimizer: A multi-objective optimization algorithm for solving engineering problems." *Applied Intelligence*, 46, no. 1 (2017), 79–95.
69. Kaveh, Ali and Vahid Reza Mahdavi. "Colliding bodies optimization: A novel meta-heuristic method." *Computers & Structures*, 139 (2014), 18–27.
70. Kaveh, Ali and Vahid Reza Mahdavi. "Multi-objective colliding bodies optimization algorithm for design of trusses." *Journal of Computational Design and Engineering*, 6, no. 1 (2019), 49–59.
71. Mirjalili, Seyedali. "The ant lion optimizer." *Advances in Engineering Software*, 83 (2015), 80–98.
72. Karmaker, Ashish Kumar, Sujit Roy, and Md Raiu Ahmed. "Analysis of the impact of electric vehicle charging station on power quality issues." In *2019 International Conference on Electrical, Computer and Communication Engineering (ECCE)* (pp. 1–6). IEEE (2019).
73. Ahmed, Abdellahi, Atif Iqbal, Irfan Khan, Abdulla Al-Wahedi, Hasan Mehrjerdi, and Syed Rahman. "Impact of EV charging station penetration on harmonic distortion level in utility distribution network: A case study of Qatar." In *2021 IEEE Texas Power and Energy Conference (TPEC)* (pp. 1–6). IEEE (2021).
74. Dharmakeerthi, C. H., N. Mithulananthan, and Tapan Kumar Saha. "Overview of the impacts of plug-in electric vehicles on the power grid." In *2011 IEEE PES Innovative Smart Grid Technologies* (pp. 1–8). IEEE (2011).

4 Comprehensive Study on Electric Vehicles

Integration with Renewable Energy, Charging Infrastructure, Model Variations, Regulatory Frameworks, and Assessing Operational Efficiency of Hybrid Electric Vehicles

Keerti Rawal and Chandra Kant

4.1 INTRODUCTION

The emission of noxious gases has a substantial adverse effect on the stability of the ecosystem. The transportation industry is responsible for a quarter of all carbon emissions and over half of the world's oil consumption [1]. At present, the electricity sector plays a crucial role in promoting a nation's economic and social stability while also protecting the environment. According to projections, the utilisation of petrol and comparable energy resources is anticipated to diminish by 50% in the coming decade. This is due to a significant surge in the adoption of electric vehicles (EVs), which will facilitate the transportation industry's transition towards a carbon-neutral economy [2, 3]. As alternatives to petrol and diesel vehicles, vehicle types such as hydrogen cars and plug-in hybrids have been developed globally in recent years. However, EVs are regarded as a superior alternative [4]. EVs have been around since the late 1800s, but the widespread availability of cheap petrol cars and their inherent restrictions stymied their development. However, with developments in electrical technology and the ongoing energy problem, EVs have regained global significance in recent decades [5]. EVs have a distinct advantage over traditional vehicles, as they rely on electric motors and lithium batteries for propulsion, resulting in zero emissions and negligible environmental impact. This chapter also explores the critical subject of intelligent charging in connection with EVs [6]. Smart charging

DOI: 10.1201/9781003311829-4

allows flexible EV charging and discharging. The EV–power grid connection creates reliable, cost-effective charging settings and boosts charger energy efficiency. To achieve super-fast EV charging, EV charging periods must match internal combustion vehicle refuelling timings, and high-powered electrical equipment manufacture must increase. High-power converters and smart charging-coordinated control minimise EV charging times [7]. Smart charging reduces grid demand and improves power transmission efficiency. Solar and wind power can also help develop sustainable EV charging options. Smart charging offers frequency adjustment and auxiliary services to EV owners, reducing energy expenses by 60% [8]. EV batteries also charge EVs and store energy for grid injection. One-way or two-way EV chargers exist. Vehicle-to-grid (V2G) is when EV batteries power the grid [9], which necessitates bidirectional electric loads capable of exchanging power between the grid and the battery [10] and [11]. Grid-to-vehicle (G2V) electricity requires one-way electric chargers. G2V systems simplify control and improve reliability [11]. V2G and G2V power exchange technologies are needed for EVs to provide ancillary services. V2G EVs use the grid to improve network performance during peak hours, while G2V EVs use it to charge their batteries [12].

Figure 4.1 illustrates the schematic representation of intelligent V2G charging within power systems [13]. This chapter delves into the extensive literature surrounding EVs and their impact on various aspects of charging infrastructure.

The factors that may have an effect on public charging infrastructure for EVs are the subject of investigation in a recent study cited as [14]. The positive and negative effects of EV charging on electricity systems are analysed in [15]. In addition, the best charging strategies are the primary focus of the review [16].

Data-mining methods; medium-term, short-term, and long-term load forecasts; and centralised or decentralised charge planning are discussed in [17]. The energy storage and diesel generator EV charging solutions are highlighted in article [18].

The work by Siddhant Kumar et al., [19], explores different battery charging arrangements, extensively discussing available battery chemistries, categorisations, and materials, as well as the effects of charging speed. The authors offer valuable guidance on selecting the most appropriate battery based on specific applications [20, 21].

Within the literature, there are investigations discussed in [22] that focus on EV energy management systems aimed at mitigating fluctuations caused by wind power connections. Figure 4.2 illustrates the smooth exchange of energy between electric vehicles and the power grid using Vehicle-to-Grid (V2G) technology. Electric vehicles bolster the grid by providing surplus energy during periods of low demand

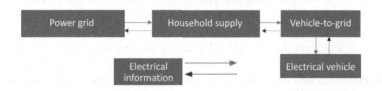

FIGURE 4.1 Vehicle-to-grid framework

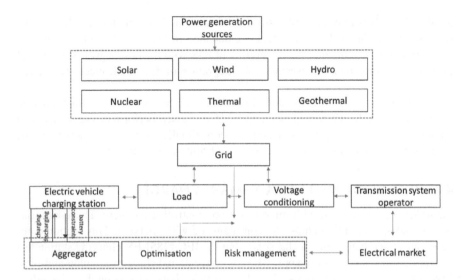

FIGURE 4.2 The diagrammatic representation of smart recharging process of vehicle-to-grid (V2G) technology integrated with power systems

and consuming electricity as needed. The dynamic interconnection improves the stability of the power grid, reduces costs, and facilitates the management of sustainable energy. Meanwhile, [23–25] concentrate on leveraging distributed photovoltaics (PVs) and wind power to reduce operational costs at EV charging stations and enhance the utilisation of renewable energy sources in station planning.

Furthermore, [26] examines EV charging methods in Germany, emphasising standards and advancements in EV technologies. The impacts of EV integration with the power grid are explored in studies presented in article [27].

Lastly, [21] provides an encompassing review of EVs as a service, addressing current challenges and exploring various applications associated with this emerging field.

Crude oil imports have surged as the road transport sector relies on imported fossil fuels. From 1998–99 to 2019, the Petroleum Planning and Analysis Cell (PPAC), Ministry of Petroleum and Natural Gas [28], reported a 5.7-fold increase in imported crude oil.

4.2 STRUCTURE OF THE ARTICLE

Section 4.3 undertakes a comparative analysis between intelligent charging and conventional charging approaches. An examination of the role and functionality of EV aggregators forms the core focus of Section 4.4. In Section 4.5, optimal planning for EV aggregators is discussed. Section 4.6 delves into the exploration of international EV charging standards, with specific attention given to the Society of Automotive Engineers (SAE) and International Electrotechnical Commission (IEC) standards. Section 4.7 provides an extensive discussion on the various types of EVs, while

Section 4.8 delves into the exploration of different models and charging levels. The research pertaining to EV converters is elucidated in Section 4.9, and Section 4.10 offers in-depth insights into the integration of EVs with the power grid. Section 4.11 provides a comprehensive overview of integration of EVs with electrical energy systems.

Distribution systems are discussed in Section 4.12. Section 4.13 introduces hybrid electric vehicles. Sections 4.14, 4.15, and 4.16 discussed performance-enhancement strategies, control strategies, and optimisation. Finally, Section 4.17 serves as the conclusion, encapsulating the key findings and summarising the article.

4.3 CONVENTIONAL VERSUS SMART CHARGE

Smart EV batteries work with microprocessor-controlled chargers. These batteries have brand-specific microchips that communicate with chargers. This section compares smart and conventional chargers using the IEC 61851 EV charging station test standard. Smart chargers use apps, whereas conventional chargers are plug-and-play. Like a conventional three-pin socket, they provide charging session data, including charging duration and kWh added. Conventional single-phase chargers can add 30 miles per hour to the EV range at 7.4 kW. They charge as fast as smart chargers.

However, traditional chargers don't know when the battery is full to cease charging. Overcharging requires manual action. Smart chargers let customers schedule charging periods, establish kWh pricing limitations, and interact with solar panels for eco-friendly charging. Smart EVs benefit from smart chargers.

Conventional chargers cannot record energy usage statistics, which can be harmful if your vehicle runs on electricity. Smart charging lets you schedule charging and discharging for the same day or in advance. Smart chargers can also enable V2G. Conventional chargers, which charge until the battery is full, are uncontrollable loads and cannot be utilised with V2G. Conventional chargers support one battery, limiting their flexibility. BEVs are zero-emission EVs according to the EPA. However, the low battery capacity of BEVs limits speed, and the lack of public charging facilities has led to the growth of plug-in hybrid electric vehicles (PHEVs) [29].

4.4 ANALYSING THE FUNCTION OF ELECTRIC VEHICLE AGGREGATORS IN INTELLIGENT CHARGING

Aggregators are new in the realm of electrical energy; they mediate between power plants and end users, giving them control over a wide range of grid-connected devices [30]. Their primary objective revolves around optimising energy programs and delivering power system control services [31]. Aggregators come in several forms, including demand response aggregators, retail aggregators, and EV aggregators [32]. An example of the role played by demand response aggregators is their provision of consumption reduction agreements to subscribers actively engaged in demand response programs [33].

The growing significance of EV integrators in power systems has generated a keen interest in the modelling of their conduct. EV aggregators act as intermediaries to enable the V2G mechanism to operate effectively. They achieve this by utilising

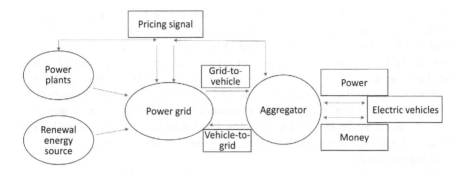

FIGURE 4.3 Exploring the involvement of EV aggregators in the implementation of smart charging

the energy capacity of EVs that are dispersed throughout the power grid [34]. They estimate the EV state-of-charge (SoC) across regions. In contracted zones, these aggregators supply electricity and manage EV charging and discharging. Figure 4.3 depicts aggregators' role in EV smart charging.

EV aggregators generate crucial signals to coordinate EV fleets. The energy market supplier, transmission system operator, and distribution system operator exchange data to coordinate.

4.5 OPTIMAL SCHEDULING FOR EV AGGREGATORS

EV aggregators must optimise scheduling. EV aggregators can optimise fleet charging and discharging schedules using complex algorithms. This optimisation considers electricity consumption, energy pricing, grid constraints, and user preferences. Optimal scheduling optimises energy use while satisfying power system constraints. This entails evaluating the appropriate charging and discharging rates for each EV in the aggregator's fleet based on battery capacity, charging infrastructure availability, and intended SoC.

EV aggregators can also engage in grid services like frequency management and peak shaving with efficient scheduling. Aggregators can maximise profitability and power grid stability by carefully altering EV charging and discharging patterns. EV aggregators use complex algorithms to schedule based on real-time data, market conditions, and system limits. These algorithms constantly monitor electricity pricing, renewable energy availability, and grid congestion to optimise charging and discharging schedules. EV aggregators can improve efficiency, efficacy, renewable energy integration, and grid stability by using optimal scheduling algorithms.

4.6 EV CHARGING STANDARDS

EV charging standards are essential for compatibility, safety, and efficiency. These standards describe EV charging station technical specs, communication protocols, and connectors, making them compatible with vehicles from different manufacturers.

SAE and IEC EV charging standards are well known. North America and Europe employ SAE standards, especially SAE J1772 and SAE Combo (CCS). The SAE Combo adds direct current (DC) rapid charging to the SAE J1772 alternating current (AC) charging port and protocol. Major cars and charging infrastructure suppliers use these standards. IEC standards, like IEC 61851, are widely accepted worldwide. IEC 61851 includes AC and DC charging modes and connectors. It specifies charging power, safety, and communication protocols. Europe uses IEC Type 2 connectors for AC and DC charging. CHAdeMO, used in Japan and backed by several international manufacturers, and GB/T, used in China, are other EV charging standards.

These standards ensure that EV owners can safely and efficiently charge their vehicles at suitable charging stations. They foster interoperability and open competition among charging infrastructure providers, enabling electric vehicle adoption.

4.7 CATEGORISATION OF EVs

When compared to vehicles powered by internal combustion engines, EVs and hybrid electric vehicles (HEVs) are cleaner and more efficient. An overview of the various EV and HEV models is provided below:

4.7.1 BATTERY ELECTRIC VEHICLES (BEVs)

BEVs are a type of vehicle that operates exclusively on electric power for propulsion. The electric motor is energised by rechargeable lithium-ion batteries that serve as an energy reservoir. Battery electric vehicles (BEVs) are devoid of an internal combustion engine and, consequently, do not emit any pollutants through the tailpipe. In order to recharge, it is necessary for them to be connected to an external power source.

4.7.2 PLUG-IN HYBRID ELECTRIC VEHICLES (PHEVs)

PHEVs integrate an internal combustion engine, an electric motor, and a rechargeable battery. These vehicles have the capability to function in either an exclusively electric mode utilising the electric motor or in a hybrid mode, where the engine and motor operate in tandem. PHEVs provide a versatile option for consumers, as they possess the capability to recharge from an external power source or, alternatively, utilise the internal combustion engine to replenish the battery. Usually, these vehicles exhibit a restricted electric-only driving distance prior to the activation of the internal combustion engine.

4.7.3 EXTENDED RANGE ELECTRIC VEHICLES (EREVs)

EREVs share similarities with PHEVs, however, they possess a greater battery capacity and a lengthier all-electric driving range. The predominant mode of operation involves the utilisation of electricity, whereby the internal combustion engine functions as a means of generating power to recharge the battery in instances where it is required. Under typical driving circumstances, the emissions of the internal combustion engine are minimised as it does not serve as the direct driving force for the wheels.

4.7.4 HYBRID ELECTRIC VEHICLES (HEVs)

HEVs employ a combination of an internal combustion engine and an electric motor to generate the necessary power to propel the vehicle. The electric motor provides supplementary power to the internal combustion engine during acceleration and recuperates energy through regenerative braking. HEVs can be recharged without the need for external power sources, as the battery is replenished through a combination of engine-generated power and regenerative braking. Compared to conventional vehicles, they offer enhanced fuel efficiency and decreased emissions.

4.7.5 FUEL CELL ELECTRIC VEHICLES (FCEVs)

FCEVs utilise hydrogen fuel cells to generate electrical energy, which subsequently propels an electric motor. The fuel cell is a device that utilises a chemical reaction between hydrogen and oxygen to produce electrical energy, while releasing solely water vapour as a byproduct. FCEVs present a superior driving range and reduced refuelling duration when compared to EVs that rely on batteries. The provision of a hydrogen refuelling infrastructure is necessary for the purpose of refuelling. The diverse range of EVs and HEVs available on the market cater to the distinct driving requirements and eco-friendly inclinations of consumers. The selection of an automobile is contingent upon various factors, including the need for driving range, accessibility to charging infrastructure, and individual preferences concerning emissions and fuel economy.

4.8 ELECTRIC CHARGING MODELS

There are several charging models or approaches for EVs that cater to different charging needs and infrastructure requirements. Here are some common charging models:

4.8.1 LEVEL 1 CHARGING

The most basic and least expedient charging technique is level 1 charging. The process entails utilising a conventional residential electrical socket operating at 120 volts of AC in conjunction with a charging cable that is commonly provided with the electric vehicle. The rate of level 1 charging is relatively low, usually yielding a range of approximately 2–5 miles per hour of charging. This charging method is advantageous for overnight charging purposes in residential settings or in situations where a higher charging capacity is unnecessary.

4.8.2 LEVEL 2 CHARGING

The charging rate of Level 2 charging is comparatively higher than that of level 1 charging. In order to charge the device, a specialised charging station or wall-mounted unit must be utilised, which must be connected to a 240 V AC power source. Level 2 charging stations are characterised by a comparatively greater power output, typically spanning from 3.3 kW to 19.2 kW. The rate of charging is subject to variation contingent upon the electric vehicle in question and the power capacity of the charging station. Level 2 charging is a prevalent feature in public charging stations, workplaces, and residential installations.

4.8.3 DC FAST CHARGING (LEVEL 3 CHARGING)

DC fast charging, commonly referred to as level 3 charging, represents the most expeditious charging technique currently accessible for electric vehicles. The system employs high-capacity charging stations that provide DC power directly to the battery of the automobile. DC fast chargers have the capability to provide power ranging from 50 kW to 350 kW, thereby enabling expeditious charging. The charging rates of electric vehicles may exhibit variability contingent upon the capabilities of the EV and the power output of the charging station. DC fast charging stations are frequently situated along highways, primary thoroughfares, and publicly accessible charging networks, facilitating extended travel and expedited charging intervals.

4.8.4 WIRELESS CHARGING

Inductive charging, commonly referred to as wireless charging, obviates the necessity for tangible cords to establish a connection between the EV and the charging station. The process entails utilising a charging pad situated on the ground in conjunction with a receiver pad affixed to the undercarriage of the electric vehicle. Wireless power transfer occurs via electromagnetic fields that are generated between two pads. The technology behind wireless charging is currently undergoing further development, and the charging rates it offers are generally inferior when compared to those of level 2 or DC fast charging. The technology provides a high level of convenience and user-friendliness, particularly in situations such as self-directed charging or charging in communal parking areas.

The aforementioned charging models offer EV proprietors the opportunity to select the most appropriate charging approach in accordance with their charging demands, travel prerequisites, and the accessibility of charging facilities. The charging velocity and convenience of these models may exhibit variability, and the expansion of the charging network and infrastructure is of paramount importance in facilitating the widespread acceptance of electric vehicles.

4.9 ELECTRIC VEHICLE POWER CONVERTERS

The power converters utilised in EVs have a pivotal function in regulating the transmission of electrical power amidst the vehicle's battery and diverse electrical constituents. The following is an evaluation of power converters for electric vehicles.

4.9.1 AC–DC POWER CONVERTERS

The AC–DC power converter is accountable for the conversion of AC from the charging source, which could be a wall outlet or charging station, into DC to facilitate the charging of the EV battery. The usual constituents of this system comprise rectifiers, filters, and voltage regulators, which are implemented to guarantee a consistent and regulated DC output. The implementation of AC–DC converters can take place either within the vehicle itself, referred to as onboard, or externally through charging infrastructure, referred to as off-board.

4.9.2 DC–DC POWER CONVERTERS

The DC–DC power converter is utilised to transform the high-voltage DC power sourced from the EV battery to lower voltage levels that are necessary for auxiliary systems, including but not limited to lighting, heating, ventilation, and infotainment systems. The implementation of this technology guarantees optimal power allocation and voltage stabilisation across the entirety of the automobile. DC–DC converters are commonly designed to be both space-efficient and energy-efficient in order to reduce energy losses that may occur during the conversion process.

4.9.2.1 DC–AC Power Inverters

The DC–AC power inverter is accountable for the conversion of DC power sourced from the electric vehicle battery into AC power, which is utilised to operate the electric motor(s) that facilitate the vehicle's propulsion. The function of the device is to transform the high-voltage DC power into the necessary frequency and voltage magnitude that is essential for the operation of the motor. DC–AC inverters are engineered to deliver substantial power output while upholding superior efficiency.

4.9.3 ONBOARD CHARGERS

The onboard charger is an AC–DC converter that is specifically designed to be integrated into electric vehicles. This converter allows for the charging of the vehicle's battery from external AC power sources. The system oversees the process of charging, encompassing the management of power input control, voltage regulation, and safety protocols. The typical design of onboard chargers facilitates the provision of support for level 1 and level 2 charging, thereby offering versatility for charging scenarios in both residential and public settings.

4.9.4 POWER ELECTRONICS AND CONTROL SYSTEMS

The incorporation of power electronics and control systems is crucial in the design of electric vehicle power converters, as they play a significant role in facilitating optimal power transfer and overall system efficacy. The aforementioned systems comprise a multitude of constituents, including semiconductor devices, such as metal-oxide-semiconductor field-effect transistor and insulated-gate bipolar transistor, which are utilised for power switching, gate drivers, control algorithms, and sensors. Sophisticated control methodologies and computational techniques are utilised to enhance the allocation of power, optimise energy utilisation, and guarantee secure functioning.

The optimisation and design of power converters for EVs prioritises factors such as thermal management, reliability, power density, and efficiency. The consistent advances in power electronics technology, semiconductor devices, and control algorithms are instrumental in enhancing the efficacy and effectiveness of power converters for electric vehicles. Furthermore, the integration of power converters with intelligent charging infrastructure and grid systems facilitates more sophisticated functionalities such as bidirectional power transfer and V2G capabilities, augmenting the comprehensive efficiency and adaptability of electric vehicle charging and utilisation.

4.10 ANALYSING THE INCORPORATION OF EVs INTO THE POWER GRID

The integration of electric transportation infrastructure has presented both advantages and disadvantages to electric power systems in contemporary times. The incorporation of EVs into the power grid is a critical factor in several domains, including the mitigation of harmonic distortions, the provision of reactive power assistance, and the regulation of peak demand. In order to facilitate the seamless integration of EVs into the power grid, it is imperative to establish a specialised regulatory entity that is specifically dedicated to overseeing EV aggregators. The aforementioned aggregators classify EVs according to the preferences of their owners with the aim of optimising opportunities within the electricity sector. The integration of energy storage systems, such as accumulators, with EVs can have a substantial impact on the electricity industry, whereas the impact of EVs alone is limited [35].

The escalating apprehension regarding climate change has stimulated an increasing curiosity in the implementation of integrated electrical energy systems that cater to the requirements of heating, cooling, and electricity. Through the process of integrating various components, these energy systems have the capability to fulfil a multitude of requirements concurrently. The integration of EVs with energy systems is of the utmost importance due to the direct impact of electric load fluctuations on the operational efficiency of energy conversion systems, such as electric motors and boilers.

Article [36] delves into the use of EVs as energy storage to balance the supply and demand curve in integrated energy systems. Furthermore, in [37], researchers examine the regional effects of EV penetration on integrated electrical and heat energy systems. EVs on the power grid could revolutionise the energy sector, improving grid stability, demand response, and renewable energy integration. EV integration also requires infrastructure, charging station placement, and charging pattern control. Smart charging, demand response, and sophisticated metering technology can address EV integration issues and maximise power grid and environmental benefits.

4.11 INTEGRATION OF EVs INTO ELECTRICAL ENERGY SYSTEMS

The V2G technology domain has undergone substantial progress, presenting unparalleled prospects for augmenting operational efficacy and enabling prosumers and consumers in the energy marketplace. Through the integration of vehicle-to-home (V2H) functionalities, individuals who possess rooftop solar power systems can effectively transition into EV-prosumers, leveraging their electric vehicles to self-supply during peak periods or instances of power outages [38]. Looking ahead to future power grids, EVs are poised to play a critical role, turning energy consumers into investors [39]. This section examines how EVs can help integrate renewable energy into electric energy infrastructure. Smart grids will manage electrical energy systems to balance production and consumption using modern communication, control, power electronics, and energy storage technologies. EVs may also absorb renewable energy and use it for mobility. Integrating EVs into energy systems requires considering charging network energy sources. EV charging alternatives include major power grid and distributed energy sources like renewables and non-renewables. This chapter will also analyse EV integration in six networks, with a focus on smart charging solutions [40].

4.12 DISTRIBUTION SYSTEMS

The advent and assimilation of EVs have had a significant influence on eco-friendly transportation, reconfiguring the sector. EVs are recognised as substantial electrical loads in the power sector, which poses both challenges and opportunities. As previously noted, electric vehicles have the capability to operate as energy storage units, providing electricity to both the power grid and end users as needed. In addition, they have the potential to function as decentralised energy sources, enhancing the resilience, dependability, and adaptability of the electrical grid [41].

Numerous investigations have been conducted to examine the incorporation of parking facilities with a significant number of electric vehicles into distribution networks, with the objective of assessing the potential advantages and consequences. In [42], researchers focused on comprehending the implications of such integration, while another research project [43] delved into the utilisation of EV parking lots within distribution systems, capitalising on selective participation in price-based load response programs. Additionally, investigations into the integration of EVs into off-grid distribution systems have established a two-way smart charging environment that enables efficient energy flow and management [44].

To optimise the coordination of EVs within distribution systems, Bharati et al. [45] proposed a framework employing a two-level V2G hierarchical optimisation approach. This framework relies on effective information exchange between EV aggregators and the network controller. By implementing this coordinated approach, strategic planning and management of EV charging can be achieved with the goal of minimising losses in the distribution network.

4.12.1 MICROGRIDS

Energy managers are currently engaged in the process of diversifying their energy production portfolio by integrating a significant proportion of renewable sources, with the objective of reducing reliance on non-renewable fossil fuels. The incorporation of sporadic renewable sources, commonly known as microgenerators, has resulted in the emergence of microgrids [46]. Intelligent charging facilitated by microgrids is instrumental in bolstering the utilisation of EVs by improving dependability, optimising energy consumption management, and providing economic benefits.

In [47], the authors emphasise the importance of renewable energy sources and EVs as essential solutions for managing energy consumption, reducing costs, and mitigating environmental impacts within microgrid systems. Moreover, [48] investigates the integration of EVs with microgrids, considering both grid-independent and grid-connected scenarios to explore avenues for increasing profits for EV owners while lowering operational expenses. Extensive research discussed in [49] centres on incentive programs offered to EV owners participating in microgrid demand response initiatives. Lastly, [50] presents comprehensive studies on designing hybrid island systems and incorporating EV parking and energy storage, with the goal of minimising operational and construction expenses while addressing various uncertainties. The research outcomes demonstrate that the integration of EV parking significantly reduces costs associated with energy storage installation.

4.12.2 RESIDENTIAL ELECTRIC SUPPLY

EVs will be connected to residential and other buildings in the next few years. The primary power grid or distributed energy systems can charge them. EV charging and discharging can reduce peak demand, lowering smart home electricity expenditures. EVs maximise smart home and network efficiency. The authors studied merging EVs with homes with wind turbines, energy storage, and combined heat and power generators to optimise energy consumption management [51, 52]. Wang et al. studied smart building–PHEV integration. Combining these new technologies improves power supply dependability and energy consumption management [53, 54].

4.13 HEVs

In an era dominated by EVs, HEVs continue to play an integral and complementary function. Despite the fact that EVs offer numerous advantages, such as zero emissions and enhanced energy efficiency, there are still obstacles that necessitate the presence of HEVs. One of the most significant limitations of EVs is their limited range and the time required for recharging. This is where HEVs come into play, as they provide drivers with extended range capabilities and eliminate range anxiety. HEVs employ both an ICE and an electric motor, enabling seamless transitions between the two power sources and ensuring continuous operation even when the electric battery is depleted. In addition, HEVs can be refuelled at standard petrol stations, eradicating the need for an extensive charging infrastructure. Incorporating HEVs alongside EVs helps meet the diverse demands of consumers by providing a wider range of options to accommodate various driving requirements. In addition, the presence of HEVs contributes to a smoother transition towards electrification by reducing reliance on fossil fuels and providing EV adopters with a familiar driving experience. In pursuit of a sustainable and efficient transportation future, the coexistence of HEVs and EVs functions as a transitional phase, bridging the gap between conventional vehicles and fully electric mobility, and highlights the need for both technologies.

The utilisation of dual power sources mandates the formulation of an energy management plan to efficiently allocate power across both sources. The objective of this approach is to reduce the usage of fuel while optimising the utilisation of power. HEVs utilise a battery as an auxiliary power supply that is charged by the ICE during vehicle operation and regenerative braking. During the course of the journey, the battery's SoC remains stable, suggesting that it is functioning in a way that sustains its charge. In contrast, PHEVs necessitate external charging from the mains, which enables the battery to be depleted to the allowable minimum level by the conclusion of the journey. PHEVs have the capability to function in a mode where the battery charge is depleted. PHEVs have the ability to alternate between charge-sustaining, charge-depletion, or hybrid modes, contingent upon the particular specifications.

4.14　PERFORMANCE-ENHANCEMENT STRATEGIES

Designing control strategies for HEVs/PHEVs poses a formidable challenge due to the intricate complexity of these vehicles. The principal objective of these tactics is to fulfil the power needs of the driver while concurrently mitigating fuel consumption, detrimental emissions, and guaranteeing optimal vehicle functionality. Achieving an optimal equilibrium between enhancing fuel efficiency and mitigating emissions necessitates a fastidious approach to control strategy. There exists a plethora of control strategies that have been suggested with the aim of improving the effectiveness of HEVs and PHEVs. This chapter offers a comprehensive analysis of strategies that were published prior to 2022, with a detailed categorisation and examination of each. This research provides a thorough examination of various control strategies, evaluating their respective advantages and limitations, thereby offering a comprehensive and integrated viewpoint.

The categorisation of strategies is presented in Figure 4.4, taking into account various factors such as system configuration complexity, computation time, solution nature (real, global, and local), and prior knowledge of driving patterns. Although a universally accepted definition for the term "structural complexity" is lacking, there exists a considerable consensus among diverse perspectives.

The domain of structural complexity pertains to the scrutiny of complexity classes, their inherent structure, and the interconnections among diverse complexity classes. Complexity classes are a collection of problems that exhibit comparable levels of complexity and can be formally defined using mathematical logic. Computation time refers to the length of time needed to carry out a computational process. A controller's robustness is determined by its ability to provide satisfactory

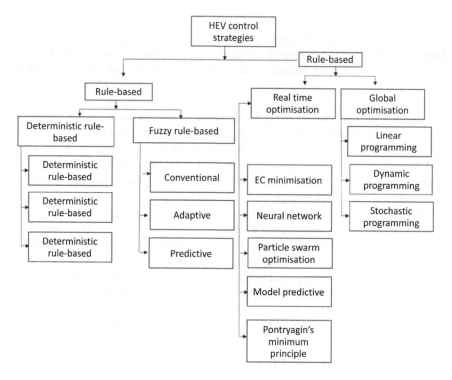

FIGURE 4.4   Categorisation of control strategies

performance in the presence of varying assumptions or uncertain parameter sets. Controllers that are robust are specifically engineered to proficiently manage sets of disturbance and uncertainty. Optimisation problems involve the identification of a local optimum, which denotes an optimal solution that is limited to a proximal set of solutions and can be either maximal or minimal. Conversely, a solution that is globally optimal represents the most optimal outcome attainable across all feasible solutions for a given optimisation problem. The categorisation of control strategies can be broadly divided into two primary groups, namely rule-based strategies and optimisation-based strategies. There exist additional subcategories that fall under these main groups.

4.15   CONTROL STRATEGIES

The classification of control strategies in hybrid drivetrains can be delineated into basic rule-based control schemes, which are executed via real-time supervisory control. The aforementioned techniques employ human knowledge, heuristics, or mathematical models to manage the flow of power, taking into account current inputs without any prior knowledge of the driving pattern. The optimisation of component operation, such as the internal combustion engine, traction motor, and generator, to meet driver demands and manage electrical loads and the battery can be achieved through the implementation of rule-based controllers. These controllers may include deterministic and fuzzy rule-based strategies. Deterministic methodologies prioritise variables such as road load to optimise fuel efficiency and enhance performance, whereas fuzzy logic controllers integrate expert knowledge and linguistic data to manage complex systems. Enhanced processing speed and memory capacity are imperative for the implementation of these control strategies, necessitating the utilisation of sophisticated microcontrollers. The utilisation of fuzzy control entails two distinct modes of operation, namely, fuel optimisation and fuzzy efficiency. The implementation of fuel-limiting control (FLC) is a strategy utilised to regulate fuel consumption and uphold the battery's SoC. Additionally, fuzzy efficiency is employed to optimise the performance of the internal combustion engine and achieve load balancing with electric motors. The utilisation of adaptive fuzzy control enables the concurrent optimisation of fuel efficiency and emissions through the allocation of weights that are based on their respective significance in varying driving contexts. The implementation of anticipatory fuzzy control techniques has the potential to mitigate fuel consumption and emissions through the analysis of GPS data, traffic conditions, and the vehicle's motion history to forecast future behaviour. In general, the utilisation of rule-based and fuzzy control strategies presents advantages such as heightened fuel efficiency, diminished emissions, enhanced drivability, and versatility in accommodating diverse driving scenarios. These benefits render such strategies valuable for energy management and overall performance in hybrid and electric vehicles.

4.16 OPTIMISATION

The principal aim of optimisation methodologies for HEVs is to augment their energy efficiency, mitigate emissions, and optimise their comprehensive performance. The design and control of HEVs involve the utilisation of various optimisation techniques. The subsequent methodologies are commonly utilised:

Global optimisation techniques are employed to ascertain the most efficient control strategy for an HEV through a comprehensive analysis of the entire driving pattern. It is deemed necessary to acquire knowledge related to various factors, including the battery's SoC, driving conditions, driver response, and route. Linear programming, dynamic programming, and genetic algorithms are examples of global optimisation techniques.

Linear programming is a mathematical methodology utilised to formulate problems related to optimising fuel efficiency in the form of convex nonlinear optimisation problems. The implementation of this methodology possesses the capability to generate outcomes that are universally optimal, specifically in the milieu of enhancing fuel efficiency in series hybrid electric automobiles.

The problem-solving methodology of dynamic programming entails the partitioning of a complicated problem into smaller subproblems, followed by the recursive determination of the most favourable solution. The present methodology exhibits the capacity to be employed in diverse scenarios, encompassing both linear and nonlinear systems, as well as constrained and unconstrained problems. However, the curse of dimensionality presents a limitation on its application in complex systems.

Genetic algorithms are a type of heuristic search methodology that employs a population-based strategy and is influenced by Darwin's theory of evolution. The procedure entails the repetitive production of innovative resolutions by means of amalgamating and altering pre-existing resolutions. Genetic algorithms have been shown to possess a high level of effectiveness in addressing complex optimisation problems that entail nonlinear and multimodal objective functions.

The focus of real-time optimisation techniques is to obtain a control strategy that minimises the cost function through the utilisation of current system variables. The implementation of an instantaneous cost function enables the facilitation of real-time energy management for HEVs.

Stochastic control strategies are employed to formulate optimisation problems that encompass variables with inherent uncertainty. Stochastic dynamic programming is employed as a means of formulating optimisation problems that extend to an infinite time horizon. Markov processes are commonly employed for predicting future power demands, and the optimal control strategy is determined by utilising stochastic dynamic programming. Stochastic control techniques are highly appropriate for optimising control policies in diverse driving patterns.

The extremum seeking technique is a methodological strategy that combines state-feedback control and stochastic dynamic programming to efficiently search for extrema. The aim is to dynamically optimise the real-time power distribution between the ICE and the electric motor. Iterative extremum seeking algorithms are utilised to continually explore the optimal solution by modifying the control inputs.

The model predictive control (MPC) approach is a control strategy that entails resolving an optimisation problem at each control step by utilising a predictive model of the system. The MPC approach considers a finite time horizon and system constraints to determine the optimal control actions. The present technology exhibits the capability to augment the contemporaneous energy administration of HEVs.

The instances mentioned above serve as examples of the optimisation techniques utilised in the design and control of HEVs. The choice of an optimisation approach for an HEV application is dependent on the specific goals, limitations, and computational capabilities available, as each technique has unique benefits and limitations.

4.17 CONCLUSIONS

The proliferation of EVs has created a need to examine and enhance charging technology and power converters to establish a versatile and dependable charging infrastructure that is economically feasible for EV batteries. The optimisation of charging efficiency is anticipated to be achieved through the utilisation of grid integration and technological innovations such as smart charging infrastructure and coordinated charging systems.

Numerous studies have been carried out pertaining to the technologies of EV chargers, as well as the global standards for EV charging, and the application standards established by prominent organisations such as IEC and SAE. This chapter examines diverse categories of electric vehicles and their assimilation into multiple energy system frameworks. The investigation of power converters utilised in EV charging systems, encompassing both AC–DC and DC–DC conversion, as well as the evaluation of EV battery performance, has been conducted. Various tiers and frameworks of EV charging have been extensively scrutinised. The utilisation of semiconductor devices and noise filters plays a pivotal role in regulating converter power in light of the broadband gap, particularly with the increasing prevalence of electric vehicles.

The implementation of novel methodologies aimed at improving power quality and grid stability is crucial for the widespread adoption of electric vehicle charging. The energy management system is a critical component of HEVs, and this research presents a comprehensive analysis of various control strategies aimed at optimising power allocation between primary and secondary sources. The methods employed for control have undergone a transformation from conventional thermostats to sophisticated intelligent techniques. The implementation of rule-based controllers is straightforward; however, it may not result in the most efficient power consumption for the entire journey. Therefore, prior knowledge of the trip is required to achieve global optimality. The optimisation-based techniques pose a challenge in achieving real-time energy management. However, a strategy based on an instantaneous cost function can facilitate a certain degree of real-time optimisation.

REFERENCES

1. S. Kumar, A. Usman, and B. S. Rajpurohit, "Battery charging topology, infrastructure, and standards for electric vehicle applications: A comprehensive review," *IET Energy Systems Integration*, vol. 3, no. 4, 2021, doi: 10.1049/esi2.12038.
2. M. Subramaniam, J. M. Solomon, V. Nadanakumar, S. Anaimuthu, and R. Sathyamurthy, "Experimental investigation on performance, combustion and emission characteristics of DI diesel engine using algae as a biodiesel," *Energy Reports*, vol. 6, 2020, doi: 10.1016/j.egyr.2020.05.022.
3. H. Wu, "A survey of battery swapping stations for electric vehicles: Operation modes and decision scenarios," *IEEE Transactions on Intelligent Transportation Systems*, vol. 23, no. 8, 2022, doi: 10.1109/TITS.2021.3125861.
4. O. Ogunkunle and N. A. Ahmed, "A review of global current scenario of biodiesel adoption and combustion in vehicular diesel engines," *Energy Reports*, vol. 5, 2019, doi: 10.1016/j.egyr.2019.10.028.
5. O. J. Oladunni, K. Mpofu, and O. A. Olanrewaju, "Greenhouse gas emissions and its driving forces in the transport sector of South Africa," *SSRN Electronic Journal*, 2021, doi: 10.2139/ssrn.3907905.
6. J. Zuo et al., "Analysis of carbon emission, carbon displacement and heterogeneity of Guangdong power industry," *Energy Reports*, vol. 8, 2022, doi: 10.1016/j.egyr.2022.03.110.
7. N. Matanov and A. Zahov, "Developments and challenges for electric vehicle charging infrastructure," 2020, doi: 10.1109/BulEF51036.2020.9326080.
8. A. Weis, P. Jaramillo, and J. Michalek, "Estimating the potential of controlled plug-in hybrid electric vehicle charging to reduce operational and capacity expansion costs for electric power systems with high wind penetration," *Applied Energy*, vol. 115, 2014, doi: 10.1016/j.apenergy.2013.10.017.
9. E. Valipour, R. Nourollahi, K. Taghizad-Tavana, S. Nojavan, and A. Alizadeh, "Risk assessment of industrial energy hubs and peer-to-peer heat and power transaction in the presence of electric vehicles," *Energies*, vol. 15, no. 23, 2022, doi: 10.3390/en15238920.
10. I. A. Umoren and M. Z. Shakir, "Electric vehicle as a service (EVaaS): Applications, challenges and enablers," *Energies*, vol. 15, no. 19, 2022, doi: 10.3390/en15197207.

11. T. He, D. D. C. Lu, M. Wu, Q. Yang, T. Li, and Q. Liu, "Four-quadrant operations of bidirectional chargers for electric vehicles in smart car parks: G2v, v2g, and v4g," *Energies*, vol. 14, no. 1, 2021, doi: 10.3390/en14010181.

12. M. Amjad, A. Ahmad, M. H. Rehmani, and T. Umer, "A review of EVs charging: From the perspective of energy optimization, optimization approaches, and charging techniques," *Transportation Research, Part D: Transport and Environment*, vol. 62, 2018, doi: 10.1016/j.trd.2018.03.006.

13. J. De Hoog et al., "Electric vehicle charging and grid constraints: Comparing distributed and centralized approaches," 2013, doi: 10.1109/PESMG.2013.6672222.

14. Q. Zhang et al., "Factors influencing the economics of public charging infrastructures for EV: A review," *Renewable and Sustainable Energy Reviews*, vol. 94, 2018, doi: 10.1016/j.rser.2018.06.022.

15. M. Nour, J. P. Chaves-Ávila, G. Magdy, and Á. Sánchez-Miralles, "Review of positive and negative impacts of electric vehicles charging on electric power systems," *Energies*, vol. 13, no. 18, 2020, doi: 10.3390/en13184675.

16. I. Rahman, P. M. Vasant, B. S. M. Singh, M. Abdullah-Al-Wadud, and N. Adnan, "Review of recent trends in optimization techniques for plug-in hybrid, and electric vehicle charging infrastructures," *Renewable and Sustainable Energy Reviews*, vol. 58, 2016, doi: 10.1016/j.rser.2015.12.353.

17. A. S. Al-Ogaili et al., "Review on scheduling, clustering, and forecasting strategies for controlling electric vehicle charging: Challenges and recommendations," *IEEE Access*, vol. 7, 2019, doi: 10.1109/ACCESS.2019.2939595.

18. A. Verma and B. Singh, "Multimode operation of solar PV array, grid, battery and diesel generator set based EV charging station," *IEEE Transactions on Industry Applications*, vol. 56, no. 5, 2020, doi: 10.1109/TIA.2020.3001268.

19. S. Chung and O. Trescases, "Hybrid energy storage system with active power-mix control in a dual-chemistry battery pack for light electric vehicles," *IEEE Transactions on Transportation Electrification*, vol. 3, no. 3, 2017, doi: 10.1109/TTE.2017.2710628.

20. A. Usman, P. Kumar, and B. P. Divakar, "Battery charging and discharging kit with DAQ to aid SOC estimation," in *4th IEEE Sponsored International Conference on Computation of Power, Energy, Information and Communication (ICCPEIC)*, pp. 13–20, Sep. 2015, doi: 10.1109/ICCPEIC.2015.7259437.

21. Y. Saleem, N. Crespi, M. H. Rehmani, and R. Copeland, "Internet of things-aided smart grid: Technologies, architectures, applications, prototypes, and future research directions," *IEEE Access*, vol. 7, pp. 62962–63003, 2019, doi: 10.1109/ACCESS.2019.2913984.

22. W. Wang, L. Liu, J. Liu, and Z. Chen, "Energy management and optimization of vehicle-to-grid systems for wind power integration," *CSEE Journal of Power and Energy Systems*, vol. 7, no. 1, 2021, doi: 10.17775/CSEEJPES.2020.01610.

23. D. Ji, M. Lv, J. Yang, and W. Yi, "Optimizing the locations and sizes of solar assisted electric vehicle charging stations in an urban area," *IEEE Access*, vol. 8, pp. 112772–112782, 2020, doi: 10.1109/ACCESS.2020.3003071.

24. A. Amer, A. Azab, M. A. Azzouz, and A. S. A. Awad, "A stochastic program for siting and sizing fast charging stations and small wind turbines in urban areas," *IEEE Transactions on Sustainable Energy*, vol. 12, no. 2, 2021, doi: 10.1109/TSTE.2020.3039910.

25. S. Shojaabadi, S. Abapour, M. Abapour, and A. Nahavandi, "Simultaneous planning of plug-in hybrid electric vehicle charging stations and wind power generation in distribution networks considering uncertainties," *Renewable Energy*, vol. 99, 2016, doi: 10.1016/j.renene.2016.06.032.

26. A. Ahmad, Z. A. Khan, M. Saad Alam, and S. Khateeb, "A review of the electric vehicle charging techniques, standards, progression and evolution of EV technologies in Germany," *Smart Science*, vol. 6, no. 1, 2018, doi: 10.1080/23080477.2017.1420132.

27. M. Singh, P. Kumar, and I. Kar, "A multi charging station for electric vehicles and its utilization for load management and the grid support," *IEEE Transactions on Smart Grid*, vol. 4, no. 2, 2013, doi: 10.1109/TSG.2013.2238562.

28. Ministry of Petroleum and Natural Gas, "Snapshot of India's oil and gas data," March 2023. https://ppac.gov.in/.

29. J. Kiviluoma and P. Meibom, "Methodology for modelling plug-in electric vehicles in the power system and cost estimates for a system with either smart or dumb electric vehicles," *Energy*, vol. 36, no. 3, 2011, doi: 10.1016/j.energy.2010.12.053.

30. S. Burger, J. P. Chaves-Ávila, C. Batlle, and I. J. Pérez-Arriaga, "A review of the value of aggregators in electricity systems," *Renewable and Sustainable Energy Reviews*, vol. 77, 2017, doi: 10.1016/j.rser.2017.04.014.

31. P. Shinde, I. Kouveliotis-Lysikatos, M. Amelin, and M. Song, "A modified progressive hedging approach for multistage intraday trade of EV aggregators," *Electric Power Systems Research*, vol. 212, 2022, doi: 10.1016/j.epsr.2022.108518.

32. L. Gkatzikis, I. Koutsopoulos, and T. Salonidis, "The role of aggregators in smart grid demand response markets," *IEEE Journal on Selected Areas in Communications*, vol. 31, no. 7, 2013, doi: 10.1109/JSAC.2013.130708.

33. M. Parvania, M. Fotuhi-Firuzabad, and M. Shahidehpour, "Optimal demand response aggregation in wholesale electricity markets," *IEEE Transactions on Smart Grid*, vol. 4, no. 4, 2013, doi: 10.1109/TSG.2013.2257894.

34. Sadeghian, O., Moradzadeh, A., Mohammadi-Ivatloo, B., Vahidinasab, V. (2022). Active Buildings Demand Response: Provision and Aggregation. In: Vahidinasab, V., Mohammadi-Ivatloo, B. (eds) Active Building Energy Systems. *Green Energy and Technology*, 355–380. Springer, Cham. https://doi.org/10.1007/978-3-030-79742-3_14.

35. M. S. Mastoi et al., "An in-depth analysis of electric vehicle charging station infrastructure, policy implications, and future trends," *Energy Reports*, vol. 8, 2022, doi: 10.1016/j.egyr.2022.09.011.

36. F. Calise, F. L. Cappiello, M. Dentice d'Accadia, and M. Vicidomini, "Smart grid energy district based on the integration of electric vehicles and combined heat and power generation," *Energy Conversion and Management*, vol. 234, p. 113932, Apr. 2021, doi: 10.1016/J.ENCONMAN.2021.113932.

37. F. Fattori, L. Tagliabue, G. Cassetti, and M. Motta, "Enhancing power system flexibility through district heating—potential role in the Italian decarbonisation," 2019, doi: 10.1109/EEEIC.2019.8783732.

38. M. S. Shamami et al., "Artificial intelligence-based performance optimization of electric vehicle-to-home (V2H) energy management system," *SAE International Journal of Sustainable Transportation, Energy, Environment, & Policy*, vol. 1, no. 2, 2020, doi: 10.4271/13-01-02-0007.

39. S. Elbatawy and W. Morsi, "Integration of prosumers with battery storage and electric vehicles via transactive energy," *IEEE Transactions on Power Delivery*, vol. 37, no. 1, 2022, doi: 10.1109/TPWRD.2021.3060922.

40. M. Yilmaz and P. T. Krein, "Review of benefits and challenges of vehicle-to-grid technology," 2012, doi: 10.1109/ECCE.2012.6342356.

41. L. Xia, I. Mareels, T. Alpcan, M. Brazil, J. De Hoog, and D. A. Thomas, "A distributed electric vehicle charging management algorithm using only local measurements," 2014, doi: 10.1109/ISGT.2014.6816420.

42. T. Chen, H. Pourbabak, Z. Liang, and W. Su, "An integrated eVoucher mechanism for flexible loads in real-time retail electricity market," *IEEE Access*, vol. 5, 2017, doi: 10.1109/ACCESS.2017.2659704.

43. M. Shafie-Khah et al., "Optimal behavior of electric vehicle parking lots as demand response aggregation agents," *IEEE Transactions on Smart Grid*, vol. 7, no. 6, 2016, doi: 10.1109/TSG.2015.2496796.

44. O. Sadeghian, M. Nazari-Heris, M. Abapour, S. S. Taheri, and K. Zare, "Improving reliability of distribution networks using plug-in electric vehicles and demand response," *Journal of Modern Power Systems and Clean Energy*, vol. 7, no. 5, pp. 1189–1199, 2019, doi: 10.1007/s40565-019-0523-8.

45. G. R. Bharati and S. Paudyal, "Coordinated control of distribution grid and electric vehicle loads," *Electric Power Systems Research*, vol. 140, 2016, doi: 10.1016/j. epsr.2016.05.031.

46. B. Ramachandran, S. K. Srivastava, and D. A. Cartes, "Intelligent power management in micro grids with EV penetration," *Expert Systems with Applications*, vol. 40, no. 16, 2013, doi: 10.1016/j.eswa.2013.06.007.

47. N. Rezaei, A. Khazali, M. Mazidi, and A. Ahmadi, "Economic energy and reserve management of renewable-based microgrids in the presence of electric vehicle aggregators: A robust optimization approach," *Energy*, vol. 201, 2020, doi: 10.1016/j. energy.2020.117629.

48. M. A. Kazemi, R. Sabzehgar, and M. Rasouli, "An optimized scheduling strategy for plugged-in electric vehicles integrated into a residential smart microgrid for both grid-tied and Islanded modes," in *2017 6th International Conference on Renewable Energy Research and Applications, ICRERA 2017*, Jan. 2017, doi: 10.1109/ ICRERA.2017.8191275.

49. P. Aliasghari, B. Mohammadi-Ivatloo, M. Alipour, M. Abapour, and K. Zare, "Optimal scheduling of plug-in electric vehicles and renewable micro-grid in energy and reserve markets considering demand response program," *Journal of Cleaner Production*, vol. 186, 2018, doi: 10.1016/j.jclepro.2018.03.058.

50. Z. Yang et al., "Robust multi-objective optimal design of islanded hybrid system with renewable and diesel sources/stationary and mobile energy storage systems," *Renewable and Sustainable Energy Reviews*, vol. 148, 2021, doi: 10.1016/j.rser.2021.111295.

51. M. E. El-Hawary, "The smart grid—State-of-the-art and future trends," *Electric Power Components and Systems*, vol. 42, no. 3–4, 2014, doi: 10.1080/15325008.2013.868558.

52. M. Tasdighi, P. J. Salamati, A. Rahimikian, and H. Ghasemi, "Energy management in a smart residential building," 2012, doi: 10.1109/EEEIC.2012.6221559.

53. K. Taghizad-Tavana, M. Ghanbari-Ghalehjoughi, N. Razzaghi-Asl, S. Nojavan, and A. Alizadeh, "An overview of the architecture of home energy management system as microgrids, automation systems, communication protocols, security, and cyber challenges," *Sustainability (Switzerland)*, vol. 14, no. 23, 2022, doi: 10.3390/su142315938.

54. H. Lin, Y. Liu, Q. Sun, R. Xiong, H. Li, and R. Wennersten, "The impact of electric vehicle penetration and charging patterns on the management of energy hub: A multi-agent system simulation," *Applied Energy*, vol. 230, 2018, doi: 10.1016/j.apenergy.2018.08.083.

5 Wireless Chargers for Electric Vehicles

Anamika Das, Ananyo Bhattacharya, and Pradip Kumar Sadhu

5.1 INTRODUCTION

Distributed energy resources (DERs) and electric vehicles (EVs) are both disruptive technologies that are rapidly transforming the energy landscape. DERs include solar photovoltaic (PV) panels, wind turbines, battery storage, and other small-scale generators that can be connected to the electric grid. EVs are battery-powered vehicles that are becoming increasingly popular as a way to reduce carbon emissions and improve air quality. The integration of DERs and EVs into the electric grid presents both opportunities and challenges. DERs and EVs can help to reduce greenhouse gas emissions, improve energy security, and enhance grid resilience. On the other hand, the variability and intermittency of these resources can make it difficult to maintain a stable grid and ensure a reliable electricity supply.

It is important to analyze and optimize network operations to optimize the integration of DERs and EVs into the electric grid. This involves developing sophisticated algorithms and models that can predict energy demand and supply, identify potential bottlenecks, and balance supply and demand in real time.

There is a lot of interest in creating new trends in the field of transportation because of the current environmental context. In this situation, electrical vehicles (EVs) are anticipated to substantially reduce greenhouse gas emissions, which will result in a healthier living environment. By 2040, India plans to introduce 31 million EVs. The Indian government has initiated the National Electric Mobility Mission (NMEM) to incorporate hybrid vehicles and EVs in a progressive attempt to enhance the energy stability and quality of life of its citizens. As is evident, EVs either fully or partially rely on batteries to function. In EVs, batteries are the primary form of storage. Wireless chargers' adaptability expands the range of circumstances in which EVs may be charged, allowing for charging not only while stationary but also while shortly halted or even running. As a consequence, it is possible to cut both the price and the weight of the battery by reducing its capacity [1, 2].

Wireless charging of EV batteries is another area of research that has the potential to create a revolution in charging methods for vehicle batteries. Wireless charging eliminates the need for cables and connectors, making it more convenient and potentially safer than conventional plug-in charging. However, there are still technical challenges to overcome, including the efficiency of wireless charging and the potential impact on the grid.

DOI: 10.1201/9781003311829-5

As optimization theory requires vast literature to be elaborately explained, in this chapter, only the major challenges for building wireless chargers for EVs, like low efficiency, voltage stress across semiconductor switches, and the effect of coil misalignment, are discussed, and appropriate topologies are suggested for mitigating the same.

5.2 BATTERIES USED IN EVs

At first, the battery type used for internal combustion (IC) engines in cars was a lead–acid battery. However, because of their poor energy density, these batteries are not recommended for use in pure EVs. ZEBRA batteries, on the other hand, function at high temperatures (270–350 °C), despite having a good energy density. Due to this limitation, the technology can only be used in cars that can run continuously to keep the working temperature constant. Recently, the market has seen extensive adoption of NiMH technology. The combination of this technology's simplicity, improved energy and power density level, cost-effectiveness, and extended lifespan make it a better option for hybrid electric vehicles (HEVs) and plug-in hybrid electric vehicles (PHEVs) despite its poor efficiency and slightly higher weight than other alternatives.

Due to their specific electrical capabilities, lithium-ion batteries are the types most commonly used in EVs. In addition, to support a higher voltage level per cell, they have an acceptable energy density and a lower self-discharging rate. Despite these benefits, it is still necessary to regulate how a lithium-ion cell charges and discharges. The battery life may be significantly impacted by (i) over-current, (ii) over-voltage, or (iii) over-charging, and these conditions can potentially result in safety problems like fires or explosions. To guarantee that the charging and discharging processes are carried out as per safety criteria, a battery management system (BMS) is integrated. The BMS keeps an eye on several battery-related data for this purpose, including voltage, state of charge (SoC), temperature, and input and output current. Because of the limitation of space, only the wireless charging of EVs will be discussed in this chapter.

5.3 CHARGING MODES IN EVs

The different charging modes of EVs treat the battery pack as a single entity that is capable of receiving internal BMS-distributed current and/or voltage. There are numerous ways to charge the batteries. Three charging levels are defined in this context by the Society of Automotive Engineers (SAE) in its Standard J1772:

Level 1	Charging via a 120 V AC plug, equivalent to a typical household plug. Thus, installing any special electrical equipment beforehand is not mandatory for charging the vehicle.
Level 2	40 A, 240 V AC plug is used for charging.
Level 3	Here, as opposed to earlier stages, the charging procedure is carried out utilizing DC rather than AC currents as the output voltage.

Additionally, certain voltage and/or current requirements may be imposed by the batteries during the charging process. Due to this, one of the following four configurations is used to complete the charging process:

- **Constant-current (CC) charging:** A steady current is given to the battery throughout the charging time. Usually, a low, continuous current helps to extend battery life. But it has the demerit of taking a long time to complete the charge. Batteries made with Li-ion and NiMH technologies use this strategy.
- **Constant-voltage (CV) charging:** This maintains a specified level of voltage on batteries that are being charged. The benefit of this is that over-voltages and the harm they cause are prevented, increasing battery life. The current is reduced in this charging mode until it reaches a set value.
- **Constant-current and constant-voltage (CC–CV) charging:** There are two stages to the charge development. An applied steady current characterizes the first phase. The maximum level of current that the battery can withstand is represented by this value, which is adjusted per the manufacturer's advice. When this happens, the voltage level rises. This charging configuration changes to a constant-voltage mode when the voltage reaches its peak. In the second phase, the current steadily declines. When the current is reduced to a specific level, the charging is considered complete.
- **Multistage constant-current (MCC) charging:** It involves charging in stages. A steady current is applied during each phase. The phase affects the value. When the voltage hits a certain level, the phase changes. The quick charging time that this mode can provide is by far its greatest benefit. It is primarily utilized in rapid chargers because of this.

5.4 BENEFITS AND LIMITATIONS OF WIRELESS POWER TRANSFER TECHNOLOGY IN CHARGING EVs

Despite various electrical as well as environmental advantages of EVs, drivers are hesitant to accept them because they feel their autonomy can be compromised by this form of transportation. Therefore, to encourage the use of EVs, new simple and user-friendly techniques are required. Wireless power transfer (WPT)-based charging is a propitious technology in this sense.

The key benefits of wireless charging of EVs are:

- **Automated operation:** Without the driver's intervention, the charge/discharge can be performed.
- **Safer charge-up of the batteries:** It is unlikely to come in contact with an electrical supply, and there is no mandate for the driver to use an electrical conductor in person. In addition, in extreme weather conditions (heavy rain or snowfall), this energy transfer is safe.
- **Dynamic charging:** WPT broadens the ways EVs can be charged so that they can be charged in a short time whether in motion or stationary. This

form of charging would allow the use of smaller EV batteries and let the vehicle use more economical electrical equipment as a side benefit.

When a WPT scheme for EV charging is developed, the following parameters are considered of main significance [2]:

(i) **Power level (kW):** This defines how long it would take for the battery to be completely charged. The higher the level of power, the less the required charging time.
(ii) **Maximum charging distance (cm):** It is necessary to design considering the maximum allowable separation between the system and the load in compliance with the specification of a standard EV to be charged.
(iii) **Power transfer efficiency (η):** The efficiency of the wireless charging unit has to be comparable to that of a plug-in system.
(iv) **Positional axis tolerance:** For a typical driver to park, the task of adjusting the horizontal and vertical position of the vehicle according to the placement of charging equipment needs to be robust.
(v) **Compact size:** A regular-sized car must be able to accommodate the charging components well.

Although inductive coupling has great potential for EV charging, the prevalent difficulties of poor efficiency and inconvenience limit commercial development. The most improved development of wireless chargers for EVs is based on magnetic resonance coupling (MRC) technology, which includes designing commercial systems for a static mode of application and prototypes for stationary and dynamic modes of applications [1–3].

Limitations: Wireless charging of electric cars has various constraints that should be considered:

(i) **Efficiency:** Wireless charging solutions are often less efficient than plug-in devices. This is due to the loss during power transmission via air. Inefficient wireless charging methods might result in longer charging times and greater expenses for the same amount of electricity.
(ii) **Cost:** Wireless charging solutions may be more costly than typical cable charging systems, both in terms of initial installation costs and continuing maintenance costs.
(iii) **Range:** The range of wireless charging is often restricted, and charging stations must be located near the car. This may restrict the flexibility and convenience of wireless charging.
(iv) **Compatibility:** Not all EVs are wireless-charging capable. Some cars need specialized charging protocols or have various charging requirements, which may not be supported by all wireless charging devices.

Overall, although wireless charging technology is an exciting breakthrough in the world of EVs, it is not without restrictions. When determining whether to invest in wireless charging technology for EVs, these constraints must be carefully considered.

5.5　WPT OPERATION MODES

Three wireless charging modes can be distinguished: static, stationary, and dynamic.

Static WPT happens when a full charge is carried out while the vehicle is scheduled to be shut off in a specific location. This is true for chargers that are put in homes or parking lots.

Quasi-dynamic or stationary charging is another variety of WPT charging. For vehicles used in public transport halted at bus/tram stations or taxi lines, this form of charging is helpful.

The dynamic mode means is when charging occurs while the vehicle is typically moving on a paved lane or running on a highway. This method of charging deals with the range anxiety of the user and encourages the use of electric vehicles on long journeys [1].

Researchers at Aligarh Muslim University elaborately discussed the need for wireless charging of HEVs including mathematical analysis and simulation results, with proper economic analysis to evaluate the feasibility of the hardware developed in the Indian context. French multinational company STMicroelectronics worked with AMU University in March 2020 to jointly develop technology for solar-powered, wireless charging stations in India [4].

5.6　BASIC SCHEMATIC OF A WIRELESS CHARGER FOR EVs

Magnetic resonance technology is the most widely used technology for wireless EV chargers, and the general design of this kind of charger is shown in Figure 5.1.

The configuration of a wireless charger is typically classified into two broad sections: the (i) primary and (ii) secondary coil sides. The primary side consists of the components connected to the primary coil. The coil is installed in the charging dock connected to the utility grid, while the secondary side consists of those components which are placed inside the vehicle. The most advanced application of wireless chargers for EVs is based on inductively coupled power transfer technology. The frequency of the utility grid is insufficient to transfer power efficiently and hence power converters are also necessary. While a resonant converter connected to the secondary coil side converts the AC from the primary side to DC suitable for charging the car battery, a resonant inverter connected to the primary coil side increases the frequency of primary voltage. To enhance power transfer efficiency, two compensation networks are further connected to the coils on each side to ensure resonant conditions. The topology of the compensating network mainly improves its ability to mitigate the effect of misalignment of coils and variations in the frequency of operation.

A time-varying current is used to activate the primary coil using magnetic resonance technology. A magnetic field is consequently produced. A voltage is induced in the secondary coil's terminals as the magnetic field links through it. The amount of magnetic field needed to induce voltage depends on the geometry and composition of the coils. Resonance is used by the system to maximize power transfer [6, 7].

This WPT's magnetic field typically operates in a frequency in the hundreds of KHz range. The corresponding international organizations are creating guidelines that suggest this parameter to be 85 KHz. The simplicity of the converters connected

(a)

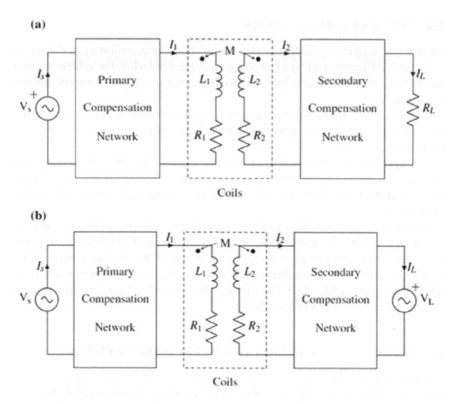

(b)

FIGURE 5.1　Basic block diagram of EV charging wirelessly; (a) unidirectional, (b) bidirectional (Source: [5])

to the secondary side is based on whether the charger is (i) unidirectional or (ii) bidirectional. The power converters on the secondary side and the battery to be charged are modeled together as a load resistance for a unidirectional charger. The charge power and voltage have an impact on the resistance value. Alternately, in a bidirectional charger, the battery is shown as a DC voltage source [5]. The equivalent circuit for the above two types of chargers is shown in Figure 5.1.

5.7　RESONANT CONVERTERS IN WIRELESS CHARGERS FOR EVs

Figure 5.2 depicts the block diagram for a resonant power converter (RPC)-based wireless charging system with six stages [8]. The source (may be a current or voltage) is the first stage. A resonant inverter, either a half- or a full-bridge type, makes up the second stage. The resonant tank circuit, which is made up of inductors and capacitors, receives power from the second stage's output. The transformer, which is the stage after that, is used to separate stages. Additionally, it can be utilized for system voltage step-up and step-down operations.

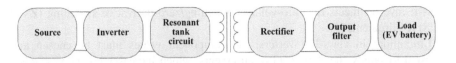

FIGURE 5.2   Block diagram of a wireless EV charging system with resonant power converters

The rectifier stage, either a full-bridge or a half-bridge rectifier, receives power from the isolation stage. The filter, which can be a low-pass filter (LPF) or a high-pass filter (HPF), receives the output of the bridge rectifier stage. The battery installed in EV is charged using the filter stage's output. Now, in the transmitter part, The MOSFET gets turned ON/OFF by a driving signal. In class-A, -B, -AB, and -C type inverters, efficiency is comparatively lower and causes a further reduction in overall transfer efficiency when used in WPTs, making them not efficient enough for resonant-coupled WPTs [2, 3, 8].

5.7.1 INVERTERS

5.7.1.1   Class-D Inverters

Class-D inverters consist of a voltage source V_{in}, two switching devices S1 and S2, an LC-series (LC-S) resonant filter with resonant frequency same as the frequency of operation, and load resistance R.

<u>Limitation</u>: MOSFET devices contain a parasitic capacitance (C_{ds}) between the drain and source. The drain-to-source voltage (V_{ds}) of a MOSFET is equal to V_{in} when it is in the off state. As the class-D type inverter contains two (2) switches, losses across the switches are stated as:

$$P_{cds} = 2 \times \frac{1}{2} C_{ds} v_s^2 f = C_{ds} v_s^2 f \qquad (5.1)$$

Thus, every instant the switch is turned ON, switching loss occurs. At low frequencies, switching losses are generally negligible. Switching losses cannot be ignored at high frequencies. One of the primary issues with class-D inverters is the deterioration in power-conversion efficiency caused by switching losses at the Megahertz frequency level.

5.7.1.2   Class-E Type Inverters

Input voltage (V_{in}), a semiconductor switch (S) with shunt capacitance (C_S), input inductance (L), a resonant filter (L-C), and load resistance (R) compose a class-E inverter. The voltages across the switch and the shunt capacitance are both zero when the switch is in the ON state, which allows current to flow through it. When the switch is in the OFF position, current passes through the shunt capacitance in the other direction. As a result, a voltage in the shape of a pulse can be seen across the switch. When the switch voltage as well as the derivative of the switch voltage are 0, the switch is turned ON. Specifically, at the instant of switch turn-on, the class-E inverter meets the zero-voltage switching/zero-derivative switching (ZVS/ZDS) requirements [5].

Limitation: In class-E inverters, the component values must be chosen and adjusted to satisfy the ZVS/ZDS requirements, which increases complexity. Also, when switching at higher switching frequencies and low resistance loads, the peak value of the voltage appearing across the switch is often between 3.5 and 4 times the V_{in}, which is unacceptable.

5.7.1.3   Class-EF$_n$ and -E/F$_n$ Inverters

In comparison with class-E inverters, class-EF$_n$ and -E/F$_n$ inverters are designed to minimize the peak stress of the voltage across switches. This may be done by deducting harmonic components from the current flowing through the C_S. A hybrid-type rectifier known as EF$_n$ or E/F$_n$, where the "n" index denotes the ratio of the resonant frequency to the switching frequency, can be created by coupling a resonant tank to F-class and F^{-1}-class inverters and integrating with E-class inverters. When "n" is even, the inverter is called an EF$_n$, and when it is odd, it is called an E/F$_n$. Low peak switch voltage and excellent efficiency are two advantages that the modified class-E inverter shares with class-E and -F inverters. The harmonic series resonant filter L_h-C_h in the class-E/F$_h$ inverter is connected in parallel with the switch of the class-E inverter.

Limitation: It is evident that the addition of a harmonic resonant filter increases the circuit volume and implementation costs.

5.7.1.4   Class-ø$_n$ Inverters

This is an extended form of the class-E/F$_n$ inverter. The class-ø$_n$ inverter shares the same circuit topology as the class-E/F$_n$ inverter. The L_h-C_h filter's resonance frequency and operating frequency are identical in the class-ø$_n$ inverter. This idea is accomplished by lowering the value of input inductance for allowing some input current to be reconverted into input voltage. The operating frequency and L_h-C_h resonant frequency are identical; therefore, the current flowing through the resonant filter stays unaffected by the load resistance. In other words, despite differences in load, the ZVS/ZDS criteria can be reached. The class-ø$_n$ inverter has this as one of its key advantages over the class-E/F inverter [5].

5.7.2   Coupling Coils

The coupling coefficient k decides whether the coils are loosely coupled or tightly coupled.

$$M = k\sqrt{L_1 L_2} \tag{5.2}$$

Where L_1 and L_2 are the self inductances of transmitting coil and receiving coil, respectively, and M is the mutual inductance between them.

As stated in the Maximum Power Transfer Theorem, the transferred power is maximum when the load impedance and the source impedance are conjugated and matched.

Thus, to achieve maximum transfer of power it is expected that:

- The imaginary part of input impedance Z_{in} equals zero [$\text{Im}(Z_{in}) = 0$].
- The real part of input impedance Z_{in} equals R_s [$\text{Re}(Z_{in}) = R_s$].

The relation between coupling coefficient k and D is expressed in the equation

$$k = \frac{1}{[1 + 2^{2/3}.D^2 / r_1 r_2]^{3/2}} \qquad (5.3)$$

where D is the lateral distance between two coupled coils, and r_1 and r_2 are the radii of the sending and receiving coil loops, respectively [9].

In the case of a two-coil (2C) system, $\text{Re}(Z_{in})$ decreases with a decrease in the value of the coupling coefficient, which means the nearer the coils, the greater the transfer efficiency. This is why a 2C system is more appropriate for transmission over a short range.

A four-coil (4C) system consists of source, transmitting, receiving, and load coils, abbreviated as SC, TC, RC, and LC, respectively. In the case of a 4C system, $\text{Re}(Z_{in})$ increases with the decrease in coupling coefficient k_{23} (coupling coefficient between TC and RC) and reaches its peak value when the gap between the TC and RC is increased up to a certain level. This is why a 4C system is more apt for mid-range transmission [3]. Figure 5.3 shows a 2C and a 4C WPT system.

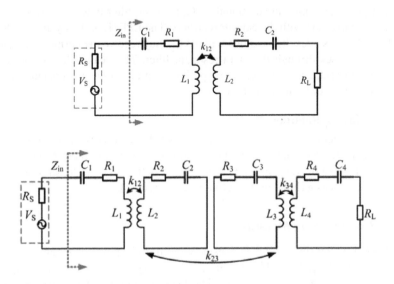

FIGURE 5.3 Two-coil and four-coil systems

5.7.3 RECTIFIERS

The receiver portion of a WPT system typically uses an AC–DC rectifier. Diodes are common switching components in rectifiers.

5.7.3.1 Class-D Type Rectifiers

The voltage across a diode is zero when it is turned on and V_o when it is turned off. As a result, the diodes experience a voltage with a rectangular form, which is known as class-D operation.

Limitations: Although theoretically, parasitic capacitance is not taken into account for a perfect diode, in practice, parasitic capacitance exists within a rectifier diode. It interferes with normal class-D operation.

5.7.3.2 Class-E Type Rectifiers

The diode D, as in the Class-D rectifier considering the shunt capacitance C_D and the LPF (L_f-C_f), comprises the class-E rectifier. When the output current is greater than the input current, the diode is turned ON and allows current flow through it. The diode is turned OFF when output current equals the input current. Shunt capacitance makes the class-E ZVS/ZDS requirements met when the diode is turned OFF. As a result, class-E rectifiers are appropriate for high-frequency (HF) applications.

Limitations: One of the disadvantages of the class-E rectifier is that, unlike the class-D rectifier, the peak voltage of the diode in the case of the class-E rectifier is significantly greater than the voltage at the output. The duty ratio of the diode (D_d) also affects the peak value.

5.7.3.3 Class-E/F$_n$ Rectifiers

Class-E/F$_n$ rectifiers have an additional L_h-C_h resonant filter tuned to harmonic frequency in comparison with the class-E rectifiers. The LPF (L_f-C_f) only allows DC to pass through it. As a result, the diode with its C_D may receive the harmonic resonant current that passes through the extra resonant filter. The peak value of the diode voltage and current can be decreased by producing a harmonic component with the appropriate phase shift and amplitude.

5.7.3.4 Class-ø$_n$ Rectifiers

A class-ø$_n$ rectifier can also be defined as "class-E/F$_n$ rectifier with zero input reactance", which slightly improves the power transfer efficiency. Input reactance can be reduced to zero by varying the Q value of the output filter. A lesser value of Q, however, introduces a ripple in the output filter current [5].

5.7.4 COMPENSATION METHODS

An air-cored transformer circuit is considered to transfer power from an AC source to a resistive load R_L. The self-inductances of primary and secondary windings are represented by L_1 and L_2 respectively. The load and circuit characteristics have a big impact on the power factor. The source's perception of the impedance changes from resistive to more inductive at relatively high frequencies. As a result, when

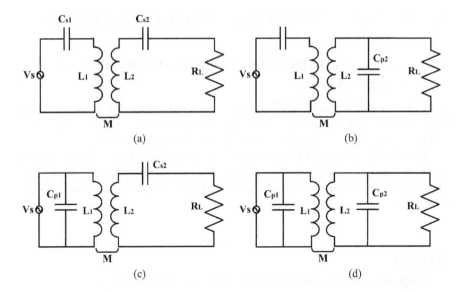

FIGURE 5.4 Basic compensation topologies (a) SS, (b) SP, (c) PS, (d) PP

the frequency rises, the power factor decreases drastically and eventually starts to approach zero. A capacitive compensating element is recommended at both primary and secondary sides to address this. (i) series–series (SS), (ii) series–parallel (SP), (iii) parallel–series (PS), and (iv) parallel–parallel (PP) are the four most common fundamental compensation topologies (Figure 5.4).

The system should run at a resonant frequency f_r to enhance power transfer capacity. Since at resonant frequency, the secondary impedance as seen from the primary is entirely resistive, the self-inductance L2 of the secondary winding is completely compensated by the secondary side compensation capacitance.

The primary side compensating circuit (capacitor in this case) is selected so that the HF inverter acting as the source has minimum volt-ampere (VA) rating, i.e., the input current and voltage are in phase [10, 11].

5.7.5 DESIGN METHODOLOGY

To achieve high power transfer efficiency, it is crucial to design a low power-loss system, taking into account the power-conversion efficiency not only of the inverter but also of the rectifier part to improve the overall efficiency [12, 13].

- First, the rectifier part is designed.
- After that, the rectifier part is expressed as the reflected impedance on the inverter side, taking into consideration the mutual inductance between coils, which depends upon distance and coil characteristics.
- Finally, the inverter part is designed to satisfy the optimal switching conditions.

5.8 DESIGN AND PERFORMANCE ANALYSIS OF EXPERIMENTAL PROTOTYPES

5.8.1 COMPARISON OF DIFFERENT COMPENSATION METHODS IN WIRELESS CHARGING CIRCUITS

A simulation model of a 2C WPT system is prepared with an HF inverter (Class-E type inverter getting input from a 100-volt DC supply with a switching frequency of 10 MHz). Table 5.1 lists the values for the parameters of the circuit

From the Simulink model based on the circuit shown in Figure 5.5(a), variation in power transfer efficiency with variation in the coupling coefficient and load resistance is recorded (Figure 5.6(a) and (b)) for four basic compensation topologies for a class-E inverter-based WPT system.

TABLE 5.1
Circuit parameters for analyzing various compensation methods

Self-Inductance of Sending Coil, L_1	Self-Inductance of Receiving Coil, L_2	Internal Resistance of Sending Coil, R_1	Internal Resistance of Receiving Coil, R_2	Compensation Capacitor at the Sending Coil Side, Cs_1	Compensation Capacitor at the Receiving Coil Side, Cs_2	Load Resistance, R_L
30 μH	30 μH	1.2 Ω	0.5 Ω	35 nF	35 nF	50 Ω

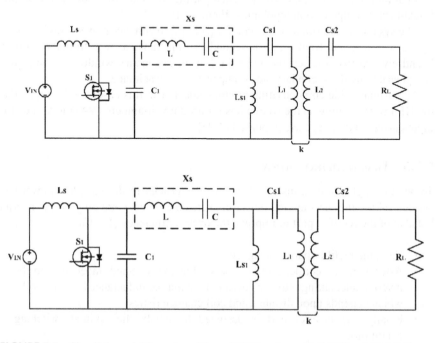

FIGURE 5.5 Simulink model for a class-E based WPT system with (a) SS and (b) LC-S type compensation

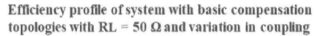

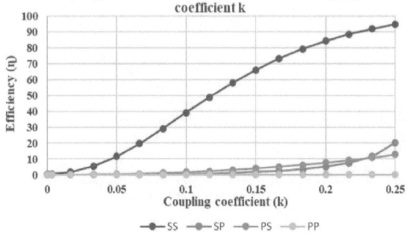

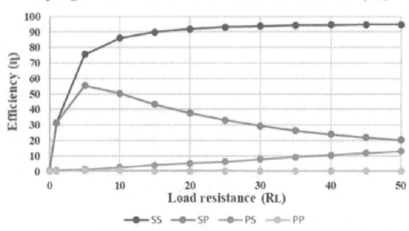

FIGURE 5.6 (a) Power transfer efficiency vs varying coupling coefficient k and (b) power transfer efficiency vs varying load resistance (R_L) for the four basic compensation topologies SS, SP, PS, and PP

A hybrid compensation topology LC-S is also proposed in which an LC compensation circuit is connected to the sending coil in addition to a capacitive compensation circuit connected in series to the receiving coil (as shown in Figure 5.5(b)). A comparison of the power transfer efficiency profile of the LC-S with SS-type compensation with variation in the coupling coefficient is recorded and shown in Figure 5.7.

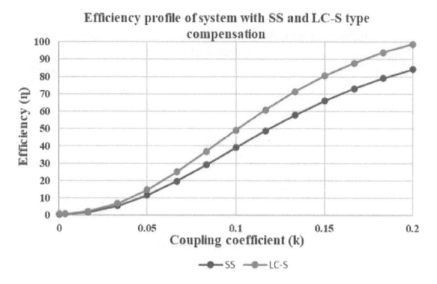

Efficiency profile of system with SS and LC-S type compensation

FIGURE 5.7   Power transfer efficiency vs varying coupling coefficient k for SS and LC-S compensation

The power transfer efficiency in four basic topologies of compensation techniques (SS, SP, PS, and PP) for WPT systems with a class-E inverter at 1 MHz switching frequency is discussed here. The comparison between different topologies shows that amongst the four, the SS topology provides the best efficiency with variation in k (R_L constant) and variation in R_L (k constant), as shown in Figure 8.6(a) and (b), respectively. In addition, the efficiency profile of LC-S type compensation is also compared and found to be a little better than that of the SS type at a fixed value of $k = 0.25$ and $R_L = 50\ \Omega$ (shown in Figure 5.7). But as can be seen from Figure 5.8, the output voltage and current profile of SS topology provide constant voltage and current in comparison with the LC-S one and is thus preferred in applications where constant voltage and current are needed, e.g., Li-ion battery charging in EVs [10].

5.9   EQUIVALENT 2C MODEL FOR A 4C WPT SYSTEM

WPT based on the inductively coupled technique may be classified into 4C and 2C structures based on their appropriateness for mid- and short-range applications, respectively. The distance between the transmitting coil and the receiving coil is crucial for an effective power transfer to the battery while charging EVs wirelessly. The 4C structure is chosen for larger spacing, while its design is more complex than that of the 2C structure. Here, a 4C structure is modified to an equivalent 2C model while keeping the same input voltage, frequency, and coupling coefficient k between the transmitting and receiving coils. The resulting model is then simulated in MATLAB®/ Simulink®.

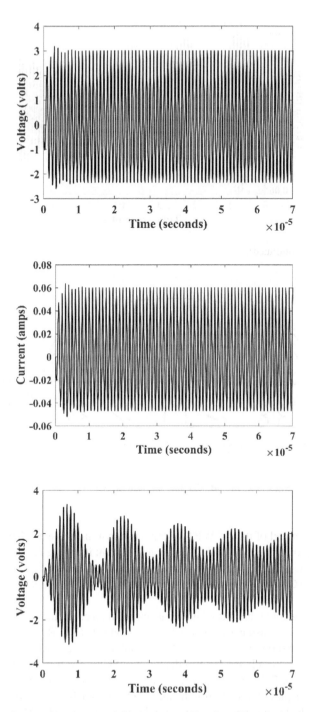

FIGURE 5.8 Output (a) voltage and (b) current profile of an SS-type compensated system. Output (c) voltage and (d) current profile of an LC-S type compensated system

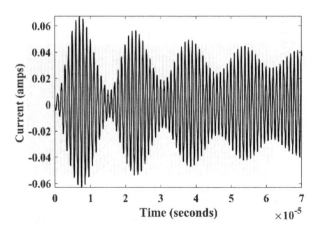

FIGURE 5.8 (Continued)

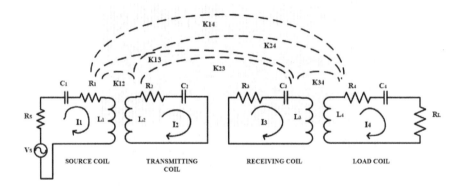

FIGURE 5.9 Circuit model of a 4C WPT system (Source: [14])

The circuit model of the 4C WPT system is shown in Figure 5.9. The source coil (SC) receives power from an HFAC source, and current is induced in the transmitting coil (TC) by the SC. The operating frequency of the TC and the receiving coil (RC) is equal. The RC, which provides power to the load, is what induces the current in the LC.

Taking into account the mutual inductances between each pair of coils, along with the internal source resistance (R_s), resistance at load side (R_L), R_1 (internal resistance of SC), R_2 (internal resistance of TC), R_3 (internal resistance of RC), and R_4 (internal resistance of LC), as well as L_1 (self-inductance of SC), L_2 (self-inductance of TC), L_3 (self-inductance of RC), and L_4 (self-inductance of LC), V_s is the HF voltage source.

Figure 5.9 displays the mutual inductances (M) for each pair of coils. Here, only the mutual inductances between pairs of adjacent coils, i.e., M_{12}, M_{23}, and M_{34}, are taken into consideration for the sake of simplicity because the flux linkages between two non-adjacent coils are too weak to be significant. The load R_L is presumptively considered entirely resistive. Next, the reflected impedance Z_{ref} between two adjacent coils is calculated.

Reflected impedance Z_{ref34} from the LC towards the RC is

$$Z_{ref34} = \frac{(\omega M_{34})^2}{Z_4}$$ (5.4)

By similarly calculating the Z_{ref23} and Z_{ref12}, overall impedance Z_{in} of the system seen from the source side is

$$Z_{in} = Z_1 + \frac{\omega^2 M_{12}^2 (Z_3 Z_4 + \omega^2 M_{34}^2)}{Z_2 Z_3 Z_4 + \omega^2 M_{34}^2 Z_2 + \omega^2 M_{23}^2 Z_4}$$ (5.5)

At resonance, the input impedance is

$$Z_{inresonance} = \text{Re}(Z_{in})$$ (5.6)

Calculated values of parameters for analyzing input characteristics are listed in Table 5.2.

Figure 5.10 shows the variation of the real component of the 4C WPT input impedance Z_{in} with the coupling coefficient.

In the 4C system, the $\text{Re}(Z_{in})$ increases as k_{23} decreases and attains its maximum value when the distance is large. The 4C scheme is hence suitable for mid-range applications.

5.9.1 EQUIVALENT 2C MODEL FOR A 4C WPT SYSTEM

A pair of two mutually coupled coils can be remodeled as a T-type arrangement of three inductors using the decoupling method. The values of these three inductors can be determined as $L_1 + M$, $L_2 + M$, and $-M$ as shown below when the dotted terminals are on the opposing sides (as illustrated in Figure 5.11).

At resonance angular frequency ω_r, values of L' and C' may be calculated from the following equations:

TABLE 5.2
Parameters values of the 4C system

Parameters	Value	Parameters	Value
f	10 MHz	k_{23}	0.1
R_1, R_4	0.1 Ω	k_{12}, k_{34}	0.7
L_1, L_4	1 μH	R_2, R_3	0.5 Ω
C_1, C_4	253pF	L_2, L_3	20 μH
R_s, R_L	50 Ω	C_2, C_3	12.67 PF

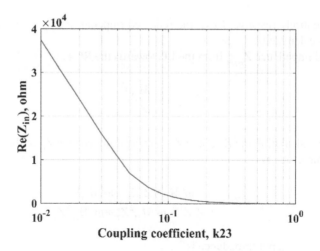

FIGURE 5.10   Variation of Re(Z_{in}) w.r.t. k_{23} in a 4C WPT system

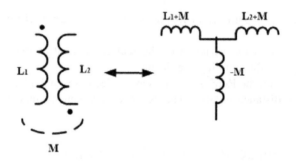

FIGURE 5.11   Equivalent T-type arrangement for two mutually coupled coils

$$L_1' = L_1 + M - \frac{1}{\omega_r^2 C_1} = k_{12}\sqrt{L_1 L_2} = M \qquad (5.7)$$

$$C_M = \frac{1}{\omega_r^2 M} \qquad (5.8)$$

$$C_2' = \frac{1}{\dfrac{1}{C_2} - \omega_r^2 M} = \frac{C_M C_2}{C_M - C_2} \qquad (5.9)$$

With the help of these equations, both transmitting- and receiving-side equivalent 2C circuits can be made (Figure 5.12).

Values of components calculated from Equations (5.7)–(5.9) for the equivalent 2C circuit are listed in Table 5.3. Figure 5.12 illustrates the corresponding 2C circuit.

TABLE 5.3

Component values of the 2C system

Components	Value	Components	value
f	10 MHz	R_2, R_3	0.5 Ω
R_S, R_L	50 Ω	C_2', C_3'	15.02 pF
C_1', C_4'	81.01 pF	L_2, L_3	20 μH
R_1, R_4	0.1 Ω	k_{12}, k_{34}	0.7
L_1', L_4'	4.13 μH	k_{23}	0.1

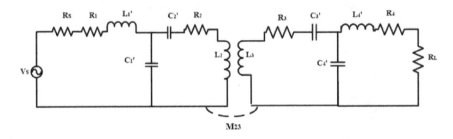

FIGURE 5.12 The LCC-type 2C equivalent circuit of a 4C WPT system

As from the 2C structure shown in Figure 5.13, the Z_{in}' may be evaluated from the equations listed below (Equations 5.10–5.14)

$$Z_4' = (R_4 + R_L + j\omega L_4')P\frac{1}{j\omega C_4}, \tag{5.10}$$

$$Z_3' = Z_4' + R_3 + \frac{1}{j\omega C_3'} \tag{5.11}$$

Considering reflected impedance from receiving side towards the transmitting side, equation for Z_{in}' yields as follows [15].

$$Z_{ref23}' = \frac{(\omega_r M_{23})^2}{Z_3'} \tag{5.12}$$

$$Z_2' = \left(Z_{ref23}' + R_2\frac{1}{j\omega_r C_2'}\right)P\frac{1}{j\omega_r C_1'} \tag{5.13}$$

$$Z_{in}' = R_1 + j\omega_r L_1' + Z_2' \tag{5.14}$$

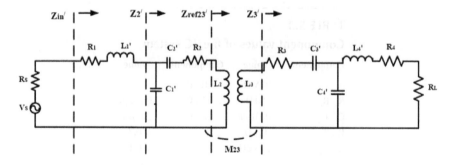

FIGURE 5.13   Circuit model for calculation of $Z_{in}{}'$ for a 2C WPT system

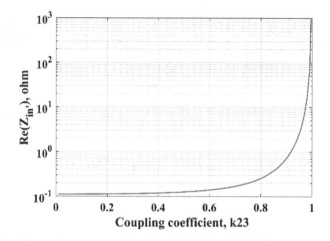

FIGURE 5.14   Variation of $Re(Z_{in}{}')$ w.r.t. k for a 2C model

Variation in $Re(Z_{in})$ in the 2C WPT with an increase in coupling coefficient (k) is shown in Figure 5.14. From this, it is evident that the smaller the spacing between coils, the better the efficiency of transmission [1]. This justifies the perception that the 2C model is more appropriate for short-range transmission [14, 16].

5.9.1.1   Simulation Results Analysis

Simulation diagrams were created in MATLAB®/Simulink® for both the 4C and its equivalent 2C LCC scheme as per the circuit in Figure 5.9 and the component values are listed in Table 5.2. The input voltage (V_{in}) applied was a sinusoidal waveform of 10 MHz frequency.

From Figures 5.15 and 5.16, it is found that the output waveform of current and voltage for a 4C scheme and its remodeled 2C counterpart are approximately

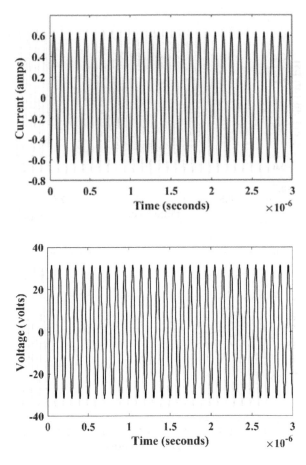

FIGURE 5.15 Instantaneous output (a) current and (b) voltage profiles of a 4C resonant WPT scheme

comparable. The comparison between output root mean square (RMS) voltage (V_{or}), output RMS current (I_{or}), output power (P_{out}), input power (P_{in}), and power transfer efficiency (η) of both the schemes are listed in Table 5.4.

In Table.5.4, it is seen that the efficiency of the 4C arrangement is 94.78%, whereas a 93.22% efficiency is noted in the corresponding 2C LCC model. Therefore, in terms of output and efficiency of power transfer, while maintaining its input impedance profile, the performance of the 2C LCC model is found to be similar to that of its corresponding 4C model [14]. Hence, it is found that the with an equivalent 2C scheme, an efficiency similar to that of a 4C scheme with larger spacing can be achieved for charging of EV batteries.

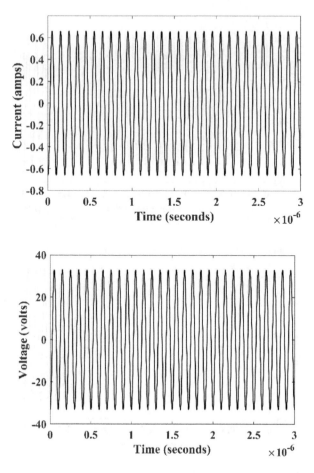

FIGURE 5.16   Instantaneous output (a) current and (b) voltage profiles of a 2C LCC resonant WPT scheme

TABLE 5.4
Comparison between 4C WPT system and its 2C equivalent

WPT model	I_{or} (Amp)	V_{or} (Volt)	P_{out} (Watts)	P_{in} (Watts)	η in percentage
4C model	0.4471	22.35	9.994	10.54	94.78
Equivalent 2C LCC model	0.4674	23.37	10.92	11.72	93.22

5.9.2   CLASS-DE RESONANT INVERTER-BASED WPT SYSTEM FOR E-SCOOTER APPLICATIONS

For WPT to achieve high power transfer efficiency, it is essential to design a resonant inductively coupled (RIC) system with low power loss, for which it is necessary to incorporate an inverter with high power-conversion efficiency. Class-DE type resonant inverters are a type of electronic power topology for energy conversion that are known for their great efficiency. The ability of the class-DE architecture to achieve both ZVS and ZDS, resulting in better efficiency in the conversion of power (DC–AC), particularly at higher switching frequencies, is an attractive characteristic. Class-DE type resonant inverters have several characteristics similar to both class-E and class-D resonant inverters.

A WPT system based on a class-DE parallel resonant voltage source inverter (VSI) is proposed in this chapter. A 100-watt system is described that can preferably be used for e-scooter applications [12]. It includes a DC source of voltage (V_{IN}), two MOSFET switches, and an L-C parallel resonant network with a capacitor C in series. The inverter output is sent to a resistive load through a pair of coupled coils. According to the suggested system, a MATLAB®/Simulink® model is created, and the output waveforms and voltage across the switches are observed.

5.9.3 Class-DE Type Parallel VSI

Figure 5.17 illustrates the class-DE type parallel resonant VSI's architecture. It contains two switches (S1 and S2), a parallel resonant Lr–Cr–R circuit, and a series capacitor (C) with a parallel resonant network. The duty ratio (D) is set to 25% for the switches S1 and S2.

The parallel resonant circuit's loaded quality factor is assumed to be high enough for the output voltage to be sinusoidal.

$$Q = \frac{R}{\omega L_r} = \omega C_r R$$

The resonant inductor is split into L_A and L_B, where $\dfrac{1}{L_r} = \dfrac{1}{L_A} + \dfrac{1}{L_B}$.

The switching frequency f is equal to the resonant frequency of the $L_A C_r$ circuit, and the phase shift of output voltage (v_o) is considered because of L_B.

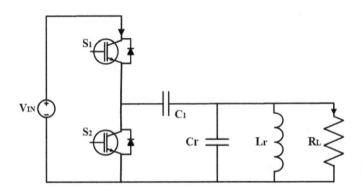

FIGURE 5.17 General circuit topology of a DE parallel resonant VSI

5.9.4 DESIGN EQUATIONS

To design the proposed circuit, considered available parameters are input DC voltage V_{in} = 100 V, operation frequency f = 85 kHz, output power P_{OUT} = 100 W, and the loaded Q factor is assumed to be equal to 5.

The equations for calculating load resistance R_L and the series capacitance C are [2]

$$R_L = \frac{V_{in}^2}{2P_{out}} \quad , \quad C = \frac{\pi}{\acute{E}\,R} \qquad (5.15)$$

Resonant inductance and capacitance can be calculated from [2]

$$L_r = \frac{R}{\acute{E}\,Q} \qquad (5.16)$$

$$C_r = \frac{1}{\acute{E}^2 L_A} = \frac{(2Q - \grave{A})}{2\acute{E}\,R} \qquad (5.17)$$

5.9.5 COMPENSATION CAPACITOR DESIGN

Compensations are added to WPT applications to improve the overall efficiency and power transmission capabilities of the system. A compensating capacitor is added to the sending and receiving side windings. The secondary (receiving) side winding compensation is used to improve the power-transfer capacity of the system, while the primary side compensation is needed to reduce the VA rating of the converter connected to the source, assuring power transmission at the unity power factor. The use of a certain compensating circuit is determined by the system's application. The SP compensation approach is utilized in this section [17, 18].

The designed parameter values are shown in Table 5.5.

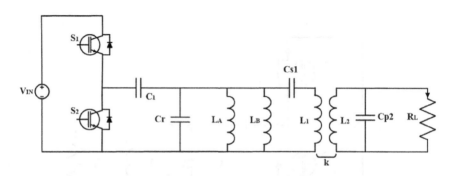

FIGURE 5.18 Proposed WPT circuit

TABLE 5.5
The designed parameter values

Parameter	Value	Parameter	Value	Parameter	Value	Parameter	Value
V_{in}	100Volts	LB	56.7 μH	C	0.12μF	r_2	0.05 Ω
P_{out}	100watts	L_1	18 μF	Lr	18.73μH	k	0.5
f	85 KHz	L_2	6 μF	Cr	0.13μF	Cs_1	0.26 μF
R	50Ω	r_1	0.04Ω	LA	26.97μH	Cp_2	0.58 μF

5.10   RESULTS AND DISCUSSION

The class-DE type parallel resonant VSI shown in Figure 5.18 was modeled in MATLAB®/Simulink®. Outputs of the simulated circuit are shown in the following figures.

From the outputs for the simulation model in MATLAB®/Simulink®, it can be seen that output voltage from the inverter exhibits a nearly sinusoidal waveform with a peak value of approximately 100 V (Figure 5.19(a)). Also, the peak voltage across switches is limited to V_{in} (Figure 5.19(b)). Figure 5.20 shows the sinusoidal (a) output voltage and (b) current respectively. The proposed WPT system simulated in MATLAB/Simulink shows a DC–AC (inversion) efficiency of 90.37%, a transmission efficiency of 81.02% over a wirelessly coupled coil with a coupling coefficient of 0.5, and overall system efficiency of 72.31%.

5.10.1   WPT System with a Push–Pull ø2 Inverter

Reduced peak voltage stress across switches, ZVS, quick transient response, and HF operation are all merits of class-ø2 inverters. In this study, the design of a WPT circuit using a class-ø2 inverter is presented. By adopting a HF resonant inverter, power density can be improved. Compared to a class-E inverter, the regulated drain-to-source terminal impedance of the switch results in soft switching and lower voltage stress. A push–pull type class-ø2 inverter with inductively coupled coils was used for WPT to boost power density and reduce input current ripple. Here, a 5.6 kW, 160 V push–pull ø2 inverter-based WPT system operating at a 30 MHz switching frequency is described. The suggested WPT system's MATLAB® simulation results are provided in this section.

5.10.2   Push–Pull Class-ø2 Inverter

Combining two single-ended inverters in parallel is the initial stage in creating a push–pull inverter, as depicted in Figure 5.10. Now, this system produces twice as much power overall as the single-ended system did. The key difference is that the two inverters' gate driving signals are 180 degrees out of phase. By sharing some components between the two inverters, the voltage and current of the other half have an

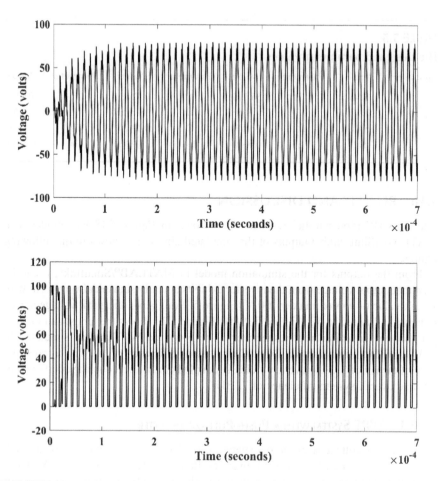

FIGURE 5.19 (a) Inverter output voltage and (b) voltage across switches

impact on one half's effective impedance. After tuning the upper inverter, a second identical copy is attached in parallel. The gate drive circuit is made to be able to generate two signals with a 180° phase shift. Two different sorts of harmonics are present in the currents produced by two inverters. Because the two inverters are activated at a 180° phase shift, the even harmonics are in phase with one another while the odd harmonics are out of phase. The upper inverter is operational for the first half. There won't be any voltage stress over the second inverter's switch at that point [13, 19, 20].

5.10.3 COMPARISON OF CLASS-E, CLASS-Ø2, AND CLASS-Ø2 PUSH–PULL INVERTER-BASED SYSTEMS

The voltage stress that appears across the switch in a class-E type inverter is known to be 3–4 times the input voltage. In comparison with class-E inverters, class-ø2

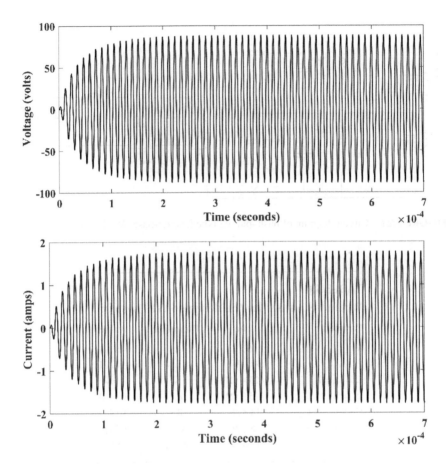

FIGURE 5.20   (a) Output voltage and (b) output current

inverters experience less voltage stress. Because of the presence of odd harmonic components, class-ø2 inverters have a very high input current ripple [20]. Using a push–pull inverter-based WPT system has several benefits, but the main one is a reduction in input harmonics content. More power may be extracted as the harmonic content is decreased. The single-ended class-ø2-based system has a power density of up to 1 kW. This push–pull design allows for the drawing of additional power.

5.11   SIMULATION RESULTS

In this part, MATLAB®/Simulink® is used to integrate a push–pull class-ø2 inverter and a full wave rectifier with appropriate compensation on both the transmitting and receiving sides (Shown in Figure 5.21). The designed parameter values are listed in Table 5.6.

It can be seen from Figure 5.22 that the input current in the push–pull topology shows lesser harmonics. Thus, the current ripple is much smaller in the proposed

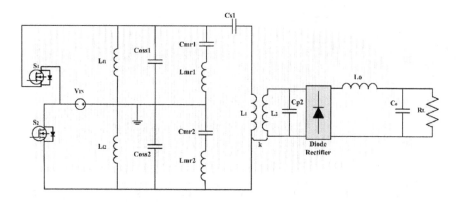

FIGURE 5.21 Circuit diagram of push–pull ø2 type inverter-based WPT system

TABLE 5.6
Parameter values for the designed WPT system

Parameter	Value	Parameter	Value
V_{input}	160 V	L_1	18 μH
f	30 MHz	L_2	18 μH
$L_f = L_{f1} = L_{f2}$	625.4 nH	k	0.4
$L_{mr} = L_{mr1} = L_{mr2}$	375.3 nH	R_{L1}	0.01 Ω
$C_{mr} = C_{mr1} = C_{mr2}$	18.8 pF	R_{L2}	0.01 Ω
$C_{oss} = C_{oss1} = C_{oss2}$	55.4 pF	C_{p_2}	1.56 pF
C_{s_1}	1.86 pF	$R_{L(load)}$	50 Ω

WPT system. which is very crucial in the design of an EV charging unit. Also, as evident from Figure 5.23, constant output voltage and current are obtained from the proposed WPT system. For $k = 0.4$, an efficiency of 67.11% with an output power of more than 6 KW is recorded.

5.12 CONCLUSION

DERs and EVs have been playing a significant role in shaping the future of the power system. DERs include small-scale power generation units such as solar panels, wind turbines, and battery storage systems, while EVs are becoming an increasingly popular mode of transportation.

The integration of DERs and EVs into the power system requires a holistic approach that takes into account the variability and uncertainty associated with these

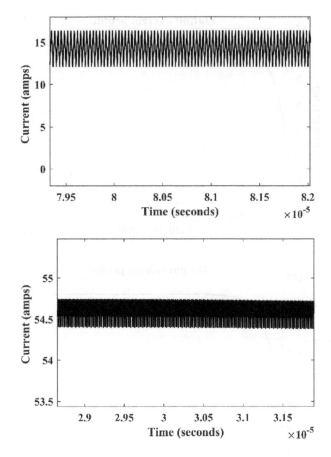

FIGURE 5.22 (a) Input current ripple of a single-ended ø2 inverter-based WPT system. (b) Input current ripple of a push–pull ø2 inverter-based WPT system

resources. The use of advanced analytical and optimization techniques can help network operators manage the challenges associated with these resources and ensure the reliability and efficient operation of the power system.

This chapter emphasizes various types of batteries used in EVs, describing their applications as well as their charging modes. Emphasizing the merit of wireless charging of batteries in EVs, it gives a synopsis of the components required and the design methodology of a wireless charging unit. The resonant converters, coils, and compensation topologies are the most crucial parts of designing a wireless charging system. Research in this field focuses on improving transfer efficiency, misalignment tolerance, and voltage stress across switches. Also, the design and performance analysis of a few experimental prototypes is discussed in detail.

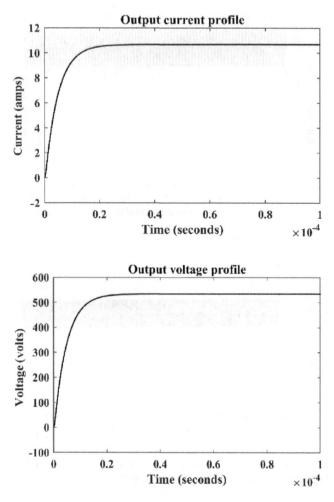

FIGURE 5.23 (a) Output current and (b) output voltage of a push–pull ø2 inverter-based WPT system

REFERENCES

1. Chirag Panchal, Sascha Stegen, and Junwei Lu, "Review of the static and dynamic wireless electric vehicle charging system," *Engineering Science and Technology, an International Journal*, 21, no. 5 (2018): 922–937.
2. M. Lu, A. Junussov, and M. Bagheri, "Analysis of resonant coupling coil configurations of EV wireless charging system: A simulation study," *Frontiers in Energy*, 14 (2020): 152–165. https://doi.org/10.1007/s11708-019-0615-1.
3. A. Triviño-Cabrera, J. M. González-González, and J. A. Aguado, "Wireless chargers for electric vehicles," in *Wireless Power Transfer for Electric Vehicles: Foundations and Design Approach. Power Systems*. Springer, Cham (2020), https://doi.org/10.1007/978-3-030-26706-3_2.

4. S. M. Shariff, M. S. Alam, F. Ahmad, Y. Rafat, M. S. J. Asghar, and S. Khan, "System design and realization of a solar-powered electric vehicle charging station," *IEEE Systems Journal*, vol. 14, no. 2 (2020): 2748–2758, doi: 10.1109/JSYST.2019.2931880.

5. Naoki Shinohara, Hidetoshi Matsuki, Hiroshi Hirayama, Hiroo Sekiya et al., "Wireless power transfer theory, technology, and applications," *IET Energy Engineering*, vol. 112, (2018): 83–109.

6. K. N. Mude and K. Aditya, "Comprehensive review and analysis of two-element resonant compensation topologies for wireless inductive power transfer systems," *Chinese Journal of Electrical Engineering*, vol. 5, no. 2 (June 2019): 14–31.

7. Chaoqiang Jiang, K. T. Chau, Chunhua Liu, and Christopher H. T. Lee, "An overview of resonant circuits for wireless power transfer," *Energies*, vol. 10, no. 7 (2017): 894.

8. Sheetal Deshmukh (Gore), Atif Iqbal, Shirazul Islam, Irfan Khan, Mousa Marzband, Syed Rahman, Abdullah M. A. B. Al-Wahedi, "Review on the classification of resonant converters for electric vehicle application," *Energy Reports*, vol. 8 (2022): 1091–1113, ISSN 2352–4847, https://doi.org/10.1016/j.egyr.2021.12.013.

9. M. Ruhul Amin and R. B. Roy, "Design and simulation of wireless stationary charging system for hybrid electric vehicle using inductive power pad in parking garage," in *The 8th International Conference on Software, Knowledge, Information Management and Applications (SKIMA)*, 2014.

10. S. Chopra and P. Bauer, "Analysis and design considerations for a contactless power transfer system," in *2011 IEEE 33rd International Telecommunications Energy Conference (INTELEC)*, pp. 1–6, 2011.

11. E. R. Joy, B. K. Kushwaha, G. Rituraj, and P. Kumar, "Analysis and comparison of four compensation topologies of contactless power transfer system," in *2015 4th International Conference on Electric Power and Energy Conversion Systems (EPECS)*, pp. 1–6, 2015, doi: 10.1109/EPECS.2015.7368544.

12. Xuezhe Wei, Zhenshi Wang, and Haifeng Dai, "A critical review of wireless power transfer via strongly coupled magnetic resonances," *Energies*, vol. 7, no. 7 (2014): 4316–4341.

13. D. Murthy-Bellur, A. Bauer, W. Kerin, and M. K. Kazimierczuk, "Inverter using loosely coupled inductors for wireless power transfer," in *2012 IEEE 55th International Midwest Symposium on Circuits and Systems (MWSCAS)*, pp. 1164–1167, 2012.

14. A. Das, A. Bhattacharya, and P. Sadhu, "Equivalent two-coil model for a four-coil wireless power transfer system," in J. Kumar and P. Jena (eds) *Recent Advances in Power Electronics and Drives. Lecture Notes in Electrical Engineering: Select Proceedings of EPREC-2020* (vol. 707). Springer, Singapore (2021). https://doi.org/10.1007/978-981-15-8586-9_11.

15. K. Sreeram, P. K. Preetha, and P. Poornachandran, "Electric vehicle scenario in India: Roadmap, challenges, and opportunities," in *2019 IEEE International Conference on Electrical, Computer and Communication Technologies (ICECCT)*, pp. 1–7, 2019, doi: 10.1109/ICECCT.2019.8869479.

16. Z. Liu, H. Zhao, C. Shuai, S. Li, "Analysis and equivalent of four-coil and two coil systems in wireless power transfer," in *2015 IEEE PELS Workshop on Emerging Technologies: Wireless Power (WoW)*, 2015.

17. T. Kondo and H. Koizumi, "Class DE voltage-source parallel resonant inverter," in *IECON 2015–41st Annual Conference of the IEEE Industrial Electronics Society*, pp. 002968–002973, 2015, doi: 10.1109/IECON.2015.7392554.

18. Anamika Das, Ananyo Bhattacharya, and Pradip Kumar Sadhu, "Class DE resonant inverter based WPT system for e-scooter applications," *Topics in Intelligent Computing And Industry Design*, vol. 3, no. 1 (2022): 23–27.

19. Juan M. Rivas, Olivia Leitermann, Yehui Han, and David J. Perreault, "A very high frequency DC–DC converter based on a class $\Phi 2$ resonant inverter," *IEEE Transactions on Power Electronics*, vol. 26, no. 10 (October 2011).
20. P. Giri, A. Das, and A. Bhattacharya, "Push-pull inverter based wireless power transfer system," in *2020 7th International Conference on Signal Processing and Integrated Networks (SPIN)*, pp. 489–493, 2020, doi: 10.1109/SPIN48934.2020.9071074.

6 Review of Control Strategy with Different Loading Conditions Considering Demand Side Management

B. Mouleeka, Sourav Chakraborty, and Susmita Kar

6.1 INTRODUCTION

The decrease in conventional energy sources leads to exploring renewable energy sources (RESs) due to their vast availability. RESs, such as wind, solar, etc., are connected to microgrids (MGs) due to their cleanliness. An MG is defined as a group of RESs, energy storage systems (ESS), and distributed energy resources (DERs) with loads that operate locally as a single controllable entity. These sources are connected to microgrid but are intermittent, which causes several power quality issues. The central console in the MG functions to maintain voltage and frequency [1]. However, in practical scenarios, increasing demand leads to an increase in generating stations, which is not feasible, and abrupt switching of loads in the distribution system is a major challenge which may lead to a state of blackout and cause overheating, reducing the system's lifespan.

MGs containing RESs (both AC and DC) are intermittent in nature, causing the continuity of power to be severely affected. Efforts have been made in the past several years to standardise these MGs [2]. However, the less inertial nature of MGs is prone to several abnormalities where unbalance in loads between phases is common in practical scenarios (distribution side). These abnormalities in the system lead to the problems in power quality such as voltage swell, voltage sag, increased lower-order harmonics, and flicker. In an islanded MG, voltage sag is crucial because medical equipment and semiconductor manufacturing automation are highly sensitive to voltage changes. In [3], the authors proposed a half-cycle discrete transformation technique for quick voltage sag detection and fast control action employing a dynamic voltage restorer (DVR). DVR is the most powerful and efficient specialised power equipment, used in contemporary power distribution networks. In [4, 5], the authors proposed DVR with a director converter which is controlled by pulse width modulation (PWM), which is generated by comparing amplitude voltage and reference voltage. In [6], the author compares DVR with proportional integral and fuzzy logic controllers. However, the results show that DVR with a fuzzy logic controller provides superior performance to mitigate system voltage dips. The author of [7] proposed the design of a second-order voltage controller with high performance. Further, the control strategy was restricted to work within 25% variation in loads, which limits its operation during peak hours. In [8], the author proposes a robust proportional resonant voltage controller incorporating negative imaginary theorem. Further, the controller provides a high level of damping and stability to the MG, which is verified with different loading and fault conditions. However, this paper lacks significant of frequency deviations. In [9], a secondary voltage controller based on reactive power-voltage (Q-V) droop control and parameter decision modular for regulation to restore voltage is proposed. However, this paper does not show significant frequency deviation in the MG. In [10], a control approach for cooperative voltage regulation is presented according to usage of decentralised sliding mode (SM) control and distributed averaging control. This control technique assumes that the q-component of each distributed generation's (DG) voltage is equal to zero and the control input's d-component is generated. However, the author must clarify reactive power sharing by adjusting the system frequency. The idea of a virtual synchronous generator (VSG), a remedy that imitates the actions of traditional synchronous generators in big power networks, has been suggested in the literature [11]. The comparison of VSGs with conventional synchronous generators is typically used to determine VSG parameters. By merging the ideas of a virtual primary, rotor, and secondary control, system frequency is stabilised/regulated. In [12] H∞ robust frequency control method is suggested for multiple microgrid. However, in multi-MG scenarios, the challenge of coordinated regulation of frequency for the AC bus of each MG is characterised as a stochastic H∞ control problem. A state feedback controller is created in a way that meets the required H∞ performance. A constraint is also implemented to assure the designed controller's logic. However, the author fails to explain the control problems for power balance with consideration of system parameters, uncertainties, and communication time delay. In [13], the authors present a decentralised AC MG frequency management that is governed by P-f droop characteristics. The active power estimation method is used to implement this strategy to avoid the use of a secondary control unit, which requires communication infrastructure. The suggested solution maintains the droop mechanism's precise power distribution while restoring the MG frequency to its nominal value by using a consensus procedure. Using the voltage setpoints of DGs and the active power droop gain, a predictive and optimal control strategy is implemented for power sharing in order to maintain the system's frequency [14]. An innovative SM control-based adaptive power point-tracking control technique is proposed to provide bidirectional fundamental frequency regulation for an MG. The SM offers an adaptive power reserve, which helps to control the change in frequency [15]. In [16] the adaptive frequency control is designed based on a data-driven algorithm to update the coefficient. Furthermore, a validation is performed to check that the coefficient is updated. In [17], the authors propose a frequency controller with an optimal distributed alternating direction method for a multiplier algorithm in which active power sharing is achieved. However, this paper lacks voltage restoration and reactive power sharing, which are mandatory for side distribution. The frequency and voltage control is mandatory for the stable operation of MGs. In

[18], the authors propose decoupled voltage and frequency controllers. Integration of a new frequency control loop to the inertia control and conventional droop control is presented to mitigate frequency. Furthermore, for voltage control, a traditional Q-V droop control method is applied. However, for situations when the penetration rate of DGs is high, the parameter values of the suggested controller must be modified to reduce the oscillations in the system, which is not feasible in practical scenarios. In [19], the authors propose a combined droop that incorporates a master–slave control strategy for the voltage and frequency control. The master VSI functions in hierarchical mode to control voltage through a droop control strategy, and the slave VSI operates in current-control mode. In [20], a novel controller is proposed with virtual impedance to control current and voltage for active and reactive power compensation. In [21], voltage and frequency deviation is mitigated through droop control and virtual impedance to regulate the loop for adjusting power distribution between DGs. High bandwidth is used for the operation of the inner control loop, which makes the system less reliable due to communication delay and packet data loss. In [22], voltage and frequency deviation is mitigated without communication by using droop control with a threshold-based feedback control path method and achieving proper power sharing. In [23], a genetic algorithm and a particle swarm optimisation algorithm are proposed to optimise the proportional integral controller for voltage and frequency control with a v–f control strategy in an MG. This method implements effective power sharing among DGs. In [24], the authors propose a proportional, integral, and derivative controller with a fuzzy proportional, integral, and derivative algorithm using an improved salp swarm algorithm for optimisation to achieve a dynamic response. In [25], the authors offer a virtual flux droop approach based on the magnitude and phase angle of virtual flux to optimise reactive and active power sharing among DG. To apply this strategy, direct flux control method is used, hence reducing the complex transformation. However, on the distribution side, the integration of single-phase loads among the phases generates a negative sequence component, which increases voltage unbalance (VU). The method proposed in [26] permits the management of both positive and negative-sequence current interactions among MGs and voltage unbalance reduction. The proposed technique avoids overcurrent stress while peak load hours on the power electronic switches, such as inverters of DERs, by focusing on the precise distribution of distinct phase currents as opposed to total active and reactive power. The suggested technique connects the DERs and MG control unit using a low-bandwidth, unidirectional communication channel. In [27], the voltage unbalance compensation is given and the dynamics are also analysed with small signal stability. This study uses the dynamic phasor method to create a comprehensive state-space model for an unbalanced MG in order to examine the dynamic behaviour. Furthermore, a small value of the integral constant leads to a slow transient response for voltage tracking. In [28], the authors propose a harmonic transfer function with $\alpha\beta$ impedance modelling, using a droop inverter in parallel for unbalanced loads to compensate for unbalance voltage. To feed unbalanced loads while maintaining balanced three-phase output voltages, the authors provide a novel grid-forming voltage control approach from a battery energy storage system (BESS). To control negative- and positive-sequence voltages and zero-sequence networks, a stationary reference frame ($\alpha\beta$)-based control scheme is proposed. It reduces the VU

at the point of common coupling [29]. In [30], the authors propose a master–slave controller for unbalance compensation based on current injection as a function of current rating to avoid overcurrent stress on inverters. In [31], the authors propose a controller for negative-sequence current sharing and negative-sequence voltage compensation in a grid-connected MG. Further voltage compensation depends on a band-pass filter, and a localised current is used for current sharing. The above references are restricted for operation during off-peak hours. However, during peak hours, when the generation and demand gap is greater, the load management is mandatory.

Several control techniques are proposed, like load shifting, valley filling, peak clipping, etc., to mitigate demand and generation mismatch [32]. The majority of the loads like ground-source heat pumps, air conditioners, and electric water heaters, are ideal candidates to take part in such a demand response (DR) method, as their energy use can be regulated without compromising consumer satisfaction. To address to the DR for peak-load shifting, one study [33] focuses on developing an optimal real-time thermal energy management system (TEMS) for intelligent homes by scheduling the heat pump for off-peak hours. The suggested TEMS combines two model-predictive controllers to monitor thermal energy storage units, a water tank, and the thermal capacity of the building. Further, the real-time pricing scheme to optimize the load usage during peak hours. The author of [34] incorporates the smart plug's functionality and investigates a low-cost, Internet of Things (IoT)-based interoperable smart plug. The design and verification of a smart plug prototype were performed through Arduino with a ZigBee-based communication protocol. However, false data injection and data theft were the major problems associated with it. In [35], three frequency control strategies, namely modified ON/OFF, droop-based, and hybrid control, are presented for household TCLs. They are simulated in two different scenarios: a TCL connected to a freestanding MG and a TCL connected to the main grid in an open loop to mitigate the gap between generation and demand. However, the heterogeneous load distribution leads to increases in VU that degrade the system performance. In [36], VU in the MG is proposed using TCLs. The lack of an inverter control strategy causes excessive usage of DSM, which leads to additional losses to the utility for a small change in loads. In [37], the authors propose a synchronised frequency control scheme for the ideal DSM solution in a stand-alone MG, where a significant number of thermostatic instruments are planned to reduce overall energy consumption during long operating hours while regulating the frequency under various conditions without compromising the quality of service (QoS). The test system under examination consists of a sluggish diesel generator, TCLs, a battery management system (BMS), and a solar photo voltaic (PV) system with a v–f control mechanism. The aggregation of loads is performed manually, which is impossible in practical scenarios. On study [38] suggests an aggregation of thermostatic loads using a space vector-based droop fed by an inverter and controlled by deep learning. Since the majority of TCLs are aggregated and scheduled during peak hours, this restores the dynamic reactive power of DGs. The author also introduces the revolutionary one-time registration of loads by incorporating an artificial neural network for simple access to customers' loads for DSM. However, the authors restrict the incorporation of DSM to only cooling loads. The authors of [39] compare the performance of particle swarm optimisation and the sine–cosine algorithm in synchronising the energy of a self-sufficient hybrid MG system with variable loads for DSM. In [40], a reinforcement learning technique is proposed for the energy trading process, modelled as a Markov decision process (MDP). Because of its superior performance in ongoing and model-free tasks, this technique was implemented to find the best course of action in the MDP. In [41], a unique methodology for residential consumers is created, employing BESS as the primary tool for centralised, incentive-based demand side control. The idea is to deploy energy storage on the premises of end customers, to support both active and reactive power. Hierarchical energy management has both primary and secondary layers of control, per [42]. In order to provide a rapid response with little communication load, semi-autonomous control engagement is selected for the majority of TCLs. The secondary control, which is centralised, seeks to reduce frequency and voltage variations by determining the optimal distribution of regulating resources, including TCL clusters. The secondary control uses a multi-objective optimisation based on trust-region power flow (PF). However, the work is restriced to temperature-dependent loads. The EV charging station also plays a major role in management during peak hours.

A fleet of electric vehicles (EVs) is used in the vehicle-to-grid (V2G) operation described in [43]; a network of charging stations connects these EVs to a distributed power system. The proposed system can respond to real-time EV usage data and identify the necessary adjustments that optimise the use of EVs to support both voltage and frequency regulation to establish V2G scheduling, taking into account the regulation prices, initial battery state of charge, EV plug-in time, battery degradation cost, preferred EV departure time, and needs of vehicle charging. To manage the charging of EV during connection sessions, the authors of [44] propose an intelligent

TABLE 6.1

Comparative table between voltage, frequency and DSM

Ref	Voltage	Frequency	DSM
[3–10]	✓		
[11–17]		✓	
[18–25]	✓	✓	
[26–31]	✓		
[32–34]			✓
[35]		✓	✓
[36]	✓		✓
[37]		✓	✓
[38]	✓		✓
[39–41]			✓
[42–44]	✓	✓	✓

Note: Tick marks refer to the covered topics in references

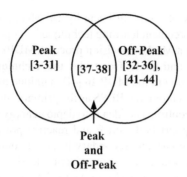

FIGURE 6.1   Circle diagram for peak and off-peak references

charging approach employing machine learning (ML) technologies. Table 6.1 shows the segregation of references for easy understanding. Figure 6.1 shows a circle diagram of references for peak and off-peak operations.

To maintain voltage and frequency, users should adhere to IEEE 1547.4. Section 6.2 illustrates the controllers, section 6.3 describes DSM, section 6.4 gives a brief discussion, and section 6.5 concludes the chapter.

6.2   CONTROL STRATEGY FOR OFF-PEAK HOURS

An MG may experience several power quality issues on the distribution side, especially due to load switching. The control action is most important to maintain nominal conditions and adhere to IEEE 1547. Several key aspects of control strategies include the following:

1) Active and reactive power should be controlled independently to regulate power flow.
2) The voltage of each DG should be updated and retained under dynamic scenarios.
3) Sharing of load should be fast (delay time is small) under isolated conditions.
4) Switching between the islanding mode and the networking mode should be smooth.

Controllers are mainly used to maintain normal conditions and provide additional security to maintain stability in the system. Deviations in voltage and frequency occur due to several dynamics in a power system. Generally, load on the distribution side follows an abnormal pattern where the requirement of additional inertia is important. In the distribution system, the three-phase load is unequally distributed at the distribution end, causing positive-sequence and negative-sequence currents in the system, which increases the voltage unbalance factor and degrades the system performance. In this regard, a feedback inverter control strategy is required that comprises voltage, frequency, and VU control. Furthermore, active and reactive power control is also required. Active power distribution in the system is based on frequency deviation, whereas reactive power distribution in the system is based on voltage deviation.

Several control methods are being adapted from past eras to mitigate voltage and frequency mismatch; these include 1) v–f control, 2) PQ control, and 3) droop control.

1) **V–f control**: voltage and frequency are fixed at the rated value in this control. This method is used to support the autonomous operation of microgrids. The pure form of v–f control cannot respond to large load changes effectively.

2) **PQ control**: Active and reactive power are fixed in this control. For grid-connected mode, PQ control is utilised such that voltage is maintained and active and reactive power remains constant in the distributed system. In this method, reactive and real power controllers are generating the reference voltage and phase angle.

3) **Droop control**: This relies on properties of drooping. In this method, voltage and frequency references are set. There are two kinds of relationships present.

a) The active power/frequency and reactive power/voltage relationships are utilised.

$$\omega = \omega_o - m_p \left(P_G - P_o \right) \tag{6.1}$$

$$V = V_o - n_q \left(Q_G - Q_o \right) \tag{6.2}$$

b) Relationships between active power/voltage and reactive power/frequency are utilised.

$$\omega = \omega_o + m_p \left(Q_G - Q_o \right) \tag{6.3}$$

$$V = V_o - n_q \left(P_G - P_o \right) \tag{6.4}$$

This method is affected by several parameters like line filter parameters, impedance parameters, and controller parameters. Here, the droop mechanism is required for power sharing among several DGs. The intermittency among the DGs puts much stress on the single operating source and hence the power sharing, as per the rating, is mandatory for continuous supply. This provides amplitude and phase control signals to the PWM and is sent to a three-phase inverter.

In MGs, different control strategies are combined with different converter control techniques. At the point of communication, there are two types of controls: centralised and de-centralised control. Centralised control relies on a communication system, whereas decentralised control relies based on local information to make decisions. Several control strategies are devised by incorporating one or more control method to maintain the normal condition in a stand-alone MG:

1) Master–slave control
2) Peer-to-peer control
3) Hierarchical control
4) SM control
5) Artificially intelligent control

6.2.1 MASTER–SLAVE CONTROL

In this control method, a strong micro-terminal is provided to maintain voltage and frequency. The source that replaces the utility grid is considered the master source, whereas the other sources are referred to as slave sources. The control strategy is heavily dependent on the master source. The capacity of the master source should be great enough in master–slave control; otherwise, it may not be able to control voltage and frequency fluctuations. Here, communication is required among micro terminals and is heavily dependent on it. By using centralised control, the load is shared among micro terminals.

6.2.2 PEER-TO-PEER CONTROL

In peer-to-peer control, each control shares the same status based on local information. This is independent of any one particular micro-source. If a fault occurs in the system, just by separating the faulty parts of the system, the MG still works properly. This control strategy does not depend on a communication system. The MG works based on local information, which is called a decentralised control strategy. Due to this, the reliability of the system increases.

6.2.3 HIERARCHICAL CONTROL

This method of control is more adaptable. Hierarchical control basically works on three levels. The fundamental layer controls the functioning of energy storage devices, loads, and micro-sources. The secondary layer controls stability and security. The tertiary layer is for communication between consumers and optimised economic performances. In this control strategy, communication levels work in two modes: centralised and decentralised.

6.2.4 SM CONTROL

In this technique, a nonlinear control induces the system to slide over a portion of its usual behaviour by employing a discontinuous control signal to change the dynamics of nonlinear systems. SM control involves:

1) Choosing a manifold or hypersurface (sliding surface) such that the system trajectory behaves well when contained within it.
2) Ensuring that the system trajectory connects and remains on the manifold, finding feedback gains.

6.2.5 ARTIFICIALLY INTELLIGENT CONTROL

This method of control is an optimisation technique that contains different algorithms like expert systems, artificial neural networks (ANNs), genetic algorithms, fuzzy logic, etc. These methods work based on a large amount of previous data present in the system as an intelligent controller.

Figure 6.2 represents the PV panels, BESS, wind, and DGs with control strategies (generation and end user side) which aggregate data based on deviations in the point of common coupling (PCC) on the generation side and through a smart meter on the load side.

6.3 DSM

DSM is a method for balancing generation and demand during peak hours. There are many techniques available for the load management system. Some of these techniques are listed here. Figure 6.3 shows DSM methods.

1) **Valley filling**: Valley filling is a method of load management to reduce peak-hour stress. The utility provider sets a variable cost for electricity, i.e., low cost during off-peak hours and high cost for peak hours, so that the customer shifts their loads to off-peak time and hence the load curve can be maintained flat.
2) **Peak clipping**: Peak clipping is a load management method that decreases coincident demand during the system's peak. Direct load control (DLC) of appliances or devices through consumer action or the use of automated controls or communications is typically employed by shifting load from the peak load period to the off-peak period.

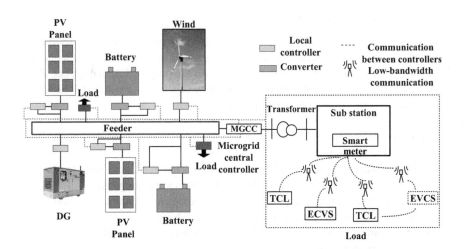

FIGURE 6.2 Single-line diagram of microgrid

3) **Load shifting**: Load shifting is one of the most prevalent methods utilised by utilities for managing peak load demand. It is based on shifting the energy consumption of consumers during peak load hours to off-peak hours of the day by providing incentives for the shifting of the load, i.e., managing the load from the demand end in order to flatten the demand curve.

4) Strategic conservation: This is implemented by utilities as part of DR initiatives to encourage consumers to alter their electricity consumption patterns. The goal is to reduce demand during peak hours and off-peak hours. It reduces both demand and total energy usage. It can be achieved by encouraging customers to use energy-efficient appliances, such as photosensitive switches for street lighting. It will reduce the load's Kvar Energy management is essential for maximising efficiency and minimising waste.

5) **Strategic load growth**: The strategic load growth strategy for load management enhances the daily extra energy consumption beyond valley filling; it is essentially a load-growing approach. It entails the implementation of novel technologies (EVs, automation, etc.) that enhance the market share of loads. It regulates a rise in the periodic use of electrical energy. For the purpose of accomplishing the strategic load growth objective, the utility will encourage smart energy systems, devices that are energy-efficient, and competitive energy sources.

6) **Flexible load figure**: Flexible load shape is mostly linked with smart power network dependability. The utility provider identifies consumers based on the type of appliances they use, either controllable or non-controllable, in order to encourage load shifting during peak hours. The consumer benefits from the utility's incentives for load shifting of controllable loads to off-peak hours to decrease stress during peak hours. The utility permits the modification of the load's form to satisfy the reliability limitations. There are customer incentives for reduced service levels. This technique incorporates entity control devices, interruptible load, and power management.

6.3.1 TCL MANIPULATION

The fundamental goal of DSM is to flatten the load curve. The usage of thermostatic loads in the distribution segment are gaining ground in recent times. During peak hour loading conditions, several control strategies are proposed to manipulate TCLs. TCLs are controlled without affecting the quality of consumer comfort. The following parameters are employed to manipulate TCL:

1) **Aggregation**: In the aggregation process, the load is clustered based on controllable and non-controllable factors.

2) **Temperature variation**: The temperature variation process aims to verify the temperature of the controllable loads which are aggregated.

3) **Segregation**: In the segregation process based on temperature variation levels, the load under certain limits or within the band range is segregated.

4) **TCL control**: The load within the band of limits is controlled with a central control smart meter.

FIGURE 6.3   DSM techniques

The following equation illustrates the functioning of TCLs.

$$\beta = \begin{cases} 1 & \theta_{Min} < \theta < \theta_{Max} \\ 0 & \theta \langle \theta_{Min}, \theta \rangle \theta_{Max} \\ \beta_{n-1} & Otherwise \end{cases} \tag{6.5}$$

Wherein,

β = Boolean variable that returns 1 or 0 to indicate whether TCLs are turned ON or OFF.

β_{n-1} = Previous state.

θ_{Max} = Maximum temperature.

θ_{Min} = Minimum temperature.

The cooling devices (ACs) and heating devices (heaters) are operated based on temperate band and scheduled according to abnormalities.

Further, electric vehicle charging stations (EVCSs) are involved in DSM. The main problem for the EVCS is the location. Various optimisation strategies have been developed for the formation of EVCS.

During peak hours, EVCS is used to supply power via V2G depending on the SOC. At night, a light load can be connected to the EVCS and charged for a prolonged period, so as cost of charging is low during the off-peak period and contributes to the system's power balance.

6.4 DISCUSSION

In an MG, unbalance is caused due to loading conditions. The loading conditions are divided into two types, peak and oof-peak hours. In the case of off-peak hours, unbalance in three phases of load leads to voltage and frequency deviations. In the distribution segment, most of the loads are single-phased, and the per-phase load

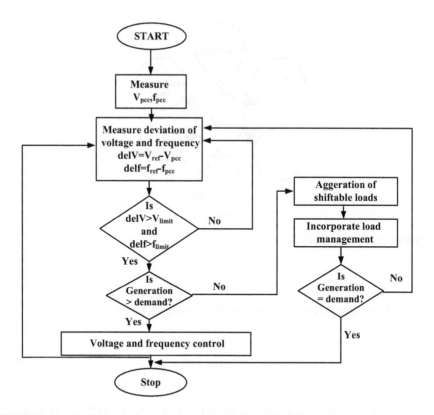

FIGURE 6.4 Block diagram of sequence control of controllers and DSM

changes in every line in such a way that per-phase voltage and current also change and eventually affect the power system characteristics. This unbalance in the system is reduced by using controllers. The unbalances in the per-phase system make it necessary for us to study the sequence components that are present in each phase individually, so each phase-sequence component is further classified as positive, negative, or zero. For a balanced system, only positive sequence components will be present, so in order to maintain stability and healthy power system characteristics, it is necessary to make the negative- and zero-sequence components equal to zero. These controllers are centralised in nature, so they are called MG central controllers (MGCCs). In the controlling part, a reference voltage is determined and the voltage of the grid is converted as positive and negative sequence components and then controlled by using the control reference frame. However, during peak hours, unbalance is caused due to the demand being more than the generation. This condition leads to the implementation of DSM. In DSM, load manipulation is conducted in many ways. The manipulation is performed in such a way to maintain both voltage and frequency stability. It can be done only during emergency conditions where the mismatch between the generation and demand is high and immediate action is necessary. Moreover, it should not affect consumers' comfort. In the DSM model, low communication bandwidth signals are used to communicate with loads. With the V2G connection of EVs, the unutilised energy in the battery will be supplied from the vehicle to the grid so as to meet the increasing demand during peak hours. Meanwhile, TCLs are controlled based on the condition of room temperature and the priority of operation. Deviation of the load is unpredictable; voltage and frequency deviation should be dealt with according to the standard of IEEE 1547.4.

6.5 CONCLUSION

This chapter focuses on control strategies under different loading conditions during peak and off-peak conditions to explain how to mitigate the stability issues in power systems that occur when non-conventional energy sources are integrated into the microgrid. During peak hours, the inverter control strategy is implemented to control voltage and frequency. However, during peak hours, the inverter control fails to mitigate the abnormalities and load end control is mandatory. Thus, TCLs scheduling and EVCS charging during peak hours is required, from which frequency deviation and voltage deviation manipulation can be performed by supplying power from unused batteries during peak hours and charging batteries again under off-peak hours. Controlling techniques are used in the system to maintain both frequency and voltage according to IEEE 1547.4.

REFERENCES

1. Lu, Yuehong, Zafar A. Khan, Manuel S. Alvarez-Alvarado, Yang Zhang, Zhijia Huang, and Muhammad Imran. "A critical review of sustainable energy policies for the promotion of renewable energy sources." *Sustainability* 12, no. 12 (2020): 5078.
2. Guerrero, Josep M., Juan C. Vasquez, José Matas, Luis García De Vicuña, and Miguel Castilla. "Hierarchical control of droop-controlled AC and DC microgrids—A general approach toward standardization." *IEEE Transactions on Industrial Electronics* 58, no. 1 (2010): 158–172.
3. Srivatchan, N. S., and P. Rangarajan. "Half cycle discrete transformation for voltage sag improvement in an islanded microgrid using dynamic voltage restorer." *International Journal of Power Electronics and Drive System* 9, no. 1 (2018): 25–32.
4. Rahman, S. Abdul, and Estifanos Dagnew. "Voltage sag compensation using direct converter based DVR by modulating the error signal." *Indonesian Journal of Electrical Engineering and Computer Science* 19, no. 2 (2020): 608–616.
5. Rahman, S. Abdul, and Gebrie Teshome. "Maximum voltage sag compensation using direct converter by modulating the carrier signal." *International Journal of Electrical and Computer Engineering* 10, no. 4 (2020): 3936.
6. Thaha, H. S., and T. R. D. Prakash, "Use of fuzzy controller based DVR for the reduction of power quality issues in composite micro-grid," In *2020 International Conference on Renewable Energy Integration into Smart Grids: A Multidisciplinary Approach to Technology Modelling and Simulation (ICREISG), Bhubaneswar, India*, pp. 131–136. IEEE, 2020, doi: 10.1109/ICREISG49226.2020.9174536.

7. Sarkar, Subrata K., Md Hassanul Karim Roni, D. Datta, Sajal K. Das, and Hemanshu R. Pota. "Improved design of high-performance controller for voltage control of islanded microgrid." *IEEE Systems Journal* 13, no. 2 (2018): 1786–1795.

8. Haque, Md Yah-Ya Ul, Md Islam, Jakir Hasan, Md Sheikh, and Rafiqul Islam. "Negative imaginary theory-based proportional resonant controller for voltage control of three-phase islanded microgrid." *Journal of Control, Automation and Electrical Systems* 32, no. 1 (2021): 214–226.

9. Xiao, Hongfei, Guangyu Liu, Jinfeng Huang, Shuaiqing Hou, and Ling Zhu. "Parameterized and centralized secondary voltage control for autonomous microgrids." *International Journal of Electrical Power & Energy Systems* 135 (2022): 107531.

10. Cucuzzella, Michele, Sebastian Trip, Antonella Ferrara, and Jacquelien Scherpen. "Cooperative voltage control in ac microgrids." In *2018 IEEE Conference on Decision and Control (CDC)*, pp. 6723–6728. IEEE, 2018.

11. Fathi, Abdulwahab, Qobad Shafiee, and Hassan Bevrani. "Robust frequency control of microgrids using an extended virtual synchronous generator." *IEEE Transactions on Power Systems* 33, no. 6 (2018): 6289–6297.

12. Hua, Haochen, Yuchao Qin, and Junwei Cao. "Coordinated frequency control for multiple microgrids in energy internet: A stochastic H∞ approach." In *2018 IEEE Innovative Smart Grid Technologies-Asia (ISGT Asia)*, pp. 810–815. IEEE, 2018.

13. Khayat, Yousef, Rasool Heydari, Mobin Naderi, Tomislav Dragicevic, Qobad Shafiee, Mohammad Fathi, Hassan Bevrani, and Frede Blaabjerg. "Estimation-based consensus approach for decentralized frequency control of ac microgrids." In *2019 21st European Conference on Power Electronics and Applications (EPE'19 ECCE Europe)*, pp. 1–8. IEEE, 2019.

14. Alghamdi, Baheej, and Claudio A. Cañizares. "Frequency regulation in isolated microgrids through optimal droop gain and voltage control." *IEEE Transactions on Smart Grid* 12, no. 2 (2020): 988–998.

15. Li, Zhongwen, Zhiping Cheng, Jikai Si, Shuyuan Zhang, Lianghui Dong, Shuhui Li, and Yixiang Gao. "Adaptive power point tracking control of PV system for primary frequency regulation of AC microgrid with high PV integration." *IEEE Transactions on Power Systems* 36, no. 4 (2021): 3129–3141.

16. Kazemi, Mohammad Verij, Seyed Jalil Sadati, and Seyed Asghar Gholamian. "Adaptive frequency control of microgrid based on fractional order control and a data-driven control with stability analysis." *IEEE Transactions on Smart Grid* 13, no. 1 (2021): 381–392.

17. Lin, Shih-Wen, and Chia-Chi Chu. "Optimal distributed ADMM-based control for frequency synchronization in isolated AC microgrids." In *IEEE Transactions on Industry Applications*, 2022.

18. Joung, Kwang Woo, Taewan Kim, and Jung-Wook Park. "Decoupled frequency and voltage control for stand-alone microgrid with high renewable penetration." *IEEE Transactions on Industry Applications* 55, no. 1 (2018): 122–133.

19. Kalke, Durvesh, H. M. Suryawanshi, Girish G. Talapur, Rohit Deshmukh, and Pratik Nachankar. "Modified droop and master-slave control for load sharing in multiple standalone AC microgrids." In *IECON 2019–45th Annual Conference of the IEEE Industrial Electronics Society*, vol. 1, pp. 1862–1867. IEEE, 2019.

20. Dashtdar, Masoud, Muhammad Shahzad Nazir, Seyed Mohammad Sadegh Hosseinimoghadam, Mohit Bajaj, and B. Srikanth Goud. "Improving the sharing of active and reactive power of the islanded microgrid based on load voltage control." *Smart Science* 10, no. 2 (2022): 142–157.

21. Heydari, Rasool, Tomislav Dragicevic, and Frede Blaabjerg. "High-bandwidth secondary voltage and frequency control of VSC-based AC microgrid." *IEEE Transactions on Power Electronics* 34, no. 11 (2019): 11320–11331.

22. Sadeque, Fahmid, and Behrooz Mirafzal. "Frequency restoration of grid-forming inverters in pulse load and plug-in events." *IEEE Journal of Emerging and Selected Topics in Industrial Electronics* 4, no. 2, (2023): 580–588, doi: 10.1109/JESTIE.2022.3186156.

23. Dashtdar, Masoud, Aymen Flah, Seyed Mohammad Sadegh Hosseinimoghadam, Ch Rami Reddy, Hossam Kotb, Kareem M. AboRas, and Edson C. Bortoni. "Improving the power quality of Island microgrid with voltage and frequency control based on a hybrid genetic algorithm and PSO." *IEEE Access* 10 (2022): 105352–105365.

24. Shayeghi, H., and A. Younesi. "Enhancement of voltage/frequency stability in an autonomous micro energy grid with penetration of wind energy using a parallel fuzzy mechanism." *Iranian Journal of Electrical and Electronic Engineering* 16, no. 4 (2020): 536–550.

25. Khanabdal, Saheb, Mahdi Banejad, Frede Blaabjerg, and Nasser Hosseinzadeh. "Adaptive virtual flux droop control based on virtual impedance in islanded AC microgrids." *IEEE Journal of Emerging and Selected Topics in Power Electronics* 10, no. 1 (2021): 1095–1107.

26. Golsorkhi, Mohammad S., David J. Hill, and Mehdi Baharizadeh. "A secondary control method for voltage unbalance compensation and accurate Load sharing in networked microgrids." *IEEE Transactions on Smart Grid* 12, no. 4 (2021): 2822–2833.

27. Peng, Yelun, Zhikang Shuai, Josep M. Guerrero, Yong Li, An Luo, and Zheng John Shen. "Performance improvement of the unbalanced voltage compensation in islanded microgrid based on small-signal analysis." *IEEE Transactions on Industrial Electronics* 67, no. 7 (2019): 5531–5542.

28. Guo, Jian, Zhiqiang Meng, Yandong Chen, Wenhua Wu, Shuhan Liao, Zhiwei Xie, and Josep M. Guerrero. "Harmonic transfer-function-based αβ-frame SISO impedance modeling of droop inverters-based islanded microgrid with unbalanced loads," *IEEE Transactions on Industrial Electronics* 70, no. 1 (2023): 452–464, doi: 10.1109/TIE.2022.3156043.

29. Xu, Bei, Victor Paduani, Hui Yu, David Lubkeman, and Ning Lu. "A novel grid-forming voltage control strategy for supplying unbalanced microgrid loads using inverter-based resources." In *2022 IEEE Power & Energy Society General Meeting (PESGM)*, pp. 1–5. IEEE, 2022.

30. Borrell, Ángel, Manel Velasco, Miguel Castilla, Jaume Miret, and Ramón Guzmán. "Collaborative voltage unbalance compensation in islanded AC microgrids with grid-forming inverters." *IEEE Transactions on Power Electronics* 37, no. 9 (2022): 10499–10513.

31. Castilla, Miguel, Manel Velasco, Jaume Miret, Ángel Borrell, and Ramón Guzmán. "Control scheme for negative-sequence voltage compensation and current sharing in inverter-based grid-connected microgrids." *IEEE Transactions on Power Electronics* 37, no. 6 (2022): 6556–6567.

32. Chakraborty, Sourav, Susmita Kar, and Deepak Kumar. "Review of demand side management with thermostatically controllable loads." In *2021 IEEE 2nd International Conference on Electrical Power and Energy Systems (ICEPES)*, pp. 1–5. IEEE, 2021.

33. Baniasadi, Ali, Daryoush Habibi, Octavian Bass, and Mohammad A. S. Masoum. "Optimal real-time residential thermal energy management for peak-load shifting with experimental verification." *IEEE Transactions on Smart Grid* 10, no. 5 (2018): 5587–5599.

34. Dhaou, Imed Ben. "Smart plug design for demand side management program." In *2019 4th International Conference on Power Electronics and their Applications (ICPEA)*, pp. 1–5. IEEE, 2019.

35. Parshin, Maksim, Maryam Majidi, Federico Ibanez, and David Pozo. "On the use of thermostatically controlled loads for frequency control." In *2019 IEEE Milan PowerTech*, pp. 1–6. IEEE, 2019.

36. Çimen, Halil, and Nurettin Çetinkaya. "Mitigation of voltage unbalance in microgrids using thermostatically controlled loads." In *2018 2nd International Symposium on Multidisciplinary Studies and Innovative Technologies (ISMSIT)*, pp. 1–4. IEEE, 2018.

37. Chakraborty, Sourav, Panneru Arvind, and Deepak Kumar. "Coordinated control for frequency regulation in a stand-alone microgrid bolstering demand side management capability." *Electric Power Components and Systems* 49, no. 1–2 (2021): 1–17.

38. Bera, Souvik, Sourav Chakraborty, Deepak Kumar, Nada Ali, and Matti Lehtonen. "Optimal deep learning based aggregation of TCLs in an inverter fed stand-alone microgrid for voltage unbalance mitigation." *Electric Power Systems Research* 210 (2022): 108178.

39. Bhuyan, Mausri, Dulal Chandra Das, and Amar Kumar Barik. "A comparative analysis of DSM based autonomous hybrid microgrid using PSO and SCA." In *2019 IEEE Region 10 Symposium (TENSYMP)*, pp. 765–770. IEEE, 2019.

40. Zhou, Suyang, Zijian Hu, Wei Gu, Meng Jiang, and Xiao-Ping Zhang. "Artificial intelligence based smart energy community management: A reinforcement learning approach." *CSEE Journal of Power and Energy Systems* 5, no. 1 (2019): 1–10.

41. Gupta, Preeti, and Yajvender Pal Verma. "Voltage profile improvement using demand side management in distribution networks under frequency linked pricing regime." *Applied Energy* 295 (2021): 117053.

42. Jendoubi, Imen, Keyhan Sheshyekani, and Hanane Dagdougui. "Aggregation and optimal management of TCLs for frequency and voltage control of a microgrid." *IEEE Transactions on Power Delivery* 36, no. 4 (2020): 2085–2096.

43. Amamra, Sid-Ali, and James Marco. "Vehicle-to-grid aggregator to support power grid and reduce electric vehicle charging cost." *IEEE Access* 7 (2019): 178528–178538.

44. López, Karol Lina, Christian Gagné, and Marc-André Gardner. "Demand-side management using deep learning for smart charging of electric vehicles." *IEEE Transactions on Smart Grid* 10, no. 3 (2018): 2683–2691.

7 Coordinated Operation of Electric Vehicle Charging Stations (EVCS) and Distributed Power Generation in Grids Using AI Technology

Mamatha N., Ramesh H.R., and Santhosha D.

7.1 INTRODUCTION

Global climate change, renewable energy sources, and electric vehicles (EVs) have been receiving a lot of attention since the 1990s. Broadly speaking, EVs with vehicle-to-grid (V2G) capabilities are ideal controllable resources that can aid distribution systems in significantly reducing peak power, regulating voltage, and offering decent stability. EVs can become more feasible resources and provide a diverse range of ancillary services by implementing V2G technologies. The operation of distribution systems is seriously influenced by this phenomenon, but there are issues with the implementation of connected power equipment, particularly resources for dispersed generation.

By leveraging cutting-edge power electronics, motor drives, energy storage technologies, renewable energy-based power generation, and smart grids, EVs, which principally run on electricity, have the potential to replace conventional internal combustion engine vehicles (ICEVs) in the near future. Ultimately, they may make a significant contribution to the decarbonization of the transportation sector. EVs can be divided into hybrid EVs (HEVs) and plug-in HEVs (PHEVs). Industrialized nations have worked hard in recent years to create a variety of financial incentives to support EV-related businesses and research initiatives. In fact, over the past ten years, infrastructure and industries related to EVs have expanded quickly. Figure 7.1 depicts the worldwide stock evolution of battery-based EVs (BEVs) and PHEVs from 2010 to 2016 to give a quick overview of the EV development trend. It is evident that the EV market has been expanding quickly and has continuously

DOI: 10.1201/9781003311829-7

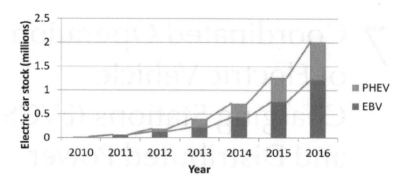

FIGURE 7.1 EV global stock evolution (Source: [2])

given researchers in related fields new technical topics to explore. The demand for building new EV charging stations (EVCSs), also known as EV supply equipment (EVSE), is anticipated to rise quickly because EVs have gradually become a part of daily life. Technical considerations, such as system configuration, power transfer rate, power and energy management, and system optimization, must be made in order to build a modern electric charging station that integrates energy storage systems (ESSs) and renewable energy (RE)-based power generation. Therefore, as is typical regarding management of RE resources and EVs, the coordinated allocation of dispersed generating resources (DGRs) and EVCSs in a V2G environment merits extensive investigation and would have considerable advantages [1]). A currently notable academic topic is how to distribute DGRs, EVCSs, and EV V2G technologies most effectively.

Regarding the ideal distribution of DGRs and EVCSs, studies have been performed from a range of perspectives and have brought attention to a plethora of application situations. Research has been done from many angles and with reference to a variety of application scenarios regarding the best distribution of DGRs and EVCSs. In order to properly allocate EVCSs, an electric power system reliability check must be included [3]). The demands of EV owners and the limitations of the functioning of the power system are thus carefully considered.

7.2 UNCOORDINATED EV CHARGING

Uncoordinated EV charging is when there is no agent or operator present and each individual EV owner has a charging schedule. Consequently, by charging their EV whenever the price of electricity is low, users can lower the cost of EV charging. The majority of unplanned EV charging takes place at home, especially in the late evening, early morning, or whenever the user gets home. Power typically flows in one direction, from the grid to the vehicles (G2V). The distribution system's effectiveness (i.e., transformer overloading, voltage instability, power loss, and frequency variation)

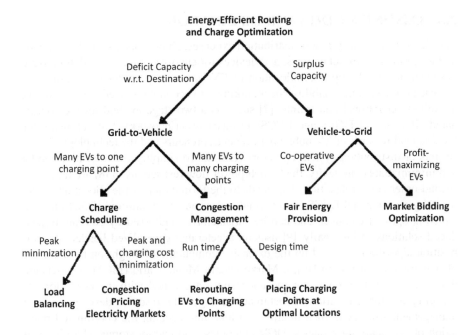

FIGURE 7.2   Representation of energy-efficient routing and charge optimization (Source: [4])

would deteriorate under a large number of recurrent, uncoordinated charges, which may cause the power system to collapse.

We give a generalized picture of the research environment in Figure 7.2 in order to conceptually make clear the connections and distinctions between the aforesaid difficulties. It is obvious that there are overlaps even when we employ a tree representation (signifying a separation between the ideas), as seen in Figure 7.2.

It is evident from this depiction of the research environment that there are a number of elements to consider depending on whether or not EVs can move based on their battery level (i.e., they need to route to their destination or charge), which in turn poses problems for G2V and V2G systems in terms of congestion control or load control, along with other things. Along with these problems, there is also the question of encouraging EV owners to use specific routes, charge at specific times (for example, to avoid peaks), or join EV communes to trade on energy markets. Regardless of the technique used to charge EVs or sell their unused capacity to the grid, infrastructure must also be constructed to manage high numbers of EVs (for example, by placing charging outlets in suitable locations).

We go into further detail on the aforementioned difficulties in the sections that follow. We propose a taxonomy of important approaches and benchmarks that may be utilized to enhance the state-of-the-art in this field by contrasting, comparing, and critically assessing these methods in order to highlight areas that require more investigation.

7.3 EXISTING MODELS AND EXPLANATIONS

In order to lower yearly power distribution expenses, DGRs such as windmills, photovoltaic modules, diesel engines, and devices for energy storage are collaboratively designed in [5]. The optimal allocation of EVCSs in [6] properly incorporates an electric power system reliability check, taking into account the needs of EV owners as well as operational constraints. [7] suggests a two-stage method for the coordinated allocation of DGRs and EVCSs in distribution networks, fully allowing for the financial rewards of EV public parking entrepreneurs and the technological constraints of distribution system operators (DSOs). To simplify the allocation model and minimize its complexity, the branch power flow restrictions in [8] are treated to second-order conic relaxation. This leads to the formulation of the allocation model for PV generation and EVCSs as a second-order cone programming (SOCP) model, which is optimally convex and can be satisfactorily addressed by commonly produced solutions. Additionally, [9] uses an accelerated generalized benders decomposition algorithm to speed up the extensive computation brought on by a variety of operation scenarios. In [10], a Markov chain Monte Carlo (MCMC) schematic diagram is used to account for the uncertainty associated with the needs for EV charging and the generation of sustainable energy. The resource development and management strategy is then modified to take into account the coordinated regulation of EV charging requests, DGR outcomes, and energy storage charging and discharging tendencies.

Similar research can be found in numerous publications regarding the V2G technologies of EVs. Concepts of unidirectional and bidirectional V2G are methodically explained in [11] and [12], respectively. Electric vehicle charging power rates and intervals are programmable in a unidirectional V2G environment to offer ancillary services. The impacts of EVs on distribution systems are correspondingly strengthened for bidirectional V2G, especially for the power demand curve, frequency control, and risk management [13]. [14] analyzes a scenario of V2G adoption inside regional power grids and employs a double-layer optimum charging approach to minimize the total load fluctuation. [15] offers a V2G-enabled EV scheduling method that is beneficial for both the customer and the aggregator from a financial standpoint.

7.4 DRAWBACKS OF THE EXISTING MODELS

The drawbacks of the existing models with electric vehicle are discussed below:

1) Inadequate charging infrastructure: A number of issues, including the high cost of EVs, particularly four-wheelers, the capacity of battery production, power consumption, charger compatibility, improper charging locations, and a lack of adequate electrical charging infrastructure, are posing problems. There are many development hurdles that the technology for charging EVs must overcome. Now let us explore the reasons why building an ecosystem for EV charging is difficult.
   a) High initial outlay: The cost of installing an EV charging station is rather significant. In order to establish the EV charging network, some

prerequisites must be met, including a suitable location, land, a vendor, electric grid stability, power restitution, a variety of charger types, connections, and other ancillary equipment. The only option to render rapid charging stations financially feasible given the high initial cost for EV charging stations is to boost their usage. The EV needs to be able to withstand high voltage or current, or both. Thus, the ideal solution for EVs may be a combination of slow and fast charges. Supercharger kits should be utilized to cut down on time further.

b) Area/location to put up the EVCS: The location of the charging point presents a significant problem when setting up an EVCS. The positioning or layout of the charging point ought to be such that it is clearly evident, conveniently accessible, and can shorten the duration of the recharging queue. In order for the consumer to reap the benefits of the most favourable moment to charge their EVs, the location should be conceived of as a great spot with qualities such as plenty of parking space, accessibility, feasibility to set up, ideal waiting area, etc.

c) Protection at EV charging stations: Sophisticated technological expertise is required while installing the EVCSs. Critical challenges include voltage variations, over-currents, frequency mismatches, and ground faults. Stabilizers, proximity sensors, and control pilot sensors must all be interconnected to prevent voltage instability. If not, it can endanger the pricey components. Many concerns, notably power problems, heat dissipation, grounding, and voltage measurement, can be addressed by adding appropriate hardware components.

d) Software-related dilemmas: One of the most crucial responsibilities is determining whether a charging spot is available. Such software is quite beneficial in this regard because it makes lives easier and improves efficiency.

In the electricity sector, for instance, there are legal limits on the use of cloud-based apps, despite the fact that they are common and essential to artificial intelligence (AI) solutions. But as the advantages of cloud applications for AI become clearer, this is changing. Second, the potential of AI in rural and other underserved parts of many developing markets, particularly low-income nations, is constrained by the reliance on cellular technology. In places where cellular network coverage is patchy or non-existent, smart meters must constantly exchange data; therefore, a lack of dependable connectivity is a major hindrance. Third, because of its digital development, the power system has become a target for hackers. The first successful strike of this sort in history took place in Ukraine in 2015, as shown in Figure 7.3, rendering thousands without electricity. Successful cyberattacks on vital infrastructure can do as much harm as a natural catastrophe. Because smart metering and automated control have grown to account for up to 10% of worldwide grid spending, or $30 billion annually devoted to digital infrastructure, the rising threat from hacking has become widespread and a topic of serious concern [**16**]. Fourth, given the diversity of the data, integrating several

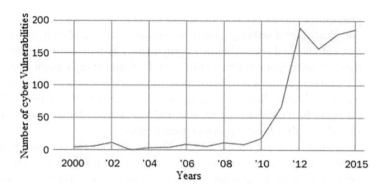

FIGURE 7.3 The number of discovered cyber vulnerabilities in industrial-control systems (Source: Kaspersky Labs. Fickling, David 2019)

data sources and guaranteeing representativeness will be difficult. Low data volumes for machine learning models to learn from may also lead to other difficulties [**17**]. Contextualizing and applying knowledge from two related activities might be challenging. Inaccurate data may also be a problem for these models. Reinforcement learning is being used in part to overcome these problems. Fifth, AI-based models pose a security concern since most of their users do not understand how they function or how they were created, making them effectively opaque. Furthermore, because current models are far from ideal, it is essential to have protections in place when integrating them into energy systems. AI may be used to automate straightforward operations, freeing up human attention for unstructured problems, when paired with improved analytics, sensors, robots, and Internet of Things devices.

e) Lengthy charging time: Chargers need to be established in regions where individuals may leave their cars parked for a considerable time because charging takes a long time. We should re-assess the charging geography in light of this. Installing charging points closer to businesses and offices will be extremely important. The government should commence imposing the obligations for setting up the infrastructure for EV charging. Encouraging global market players to provide testimonials on potential areas and significant levels of EVSE could be a starting step. The consequences of the growth in EVs on current, energy production, transmission and distribution, road traffic density, emission levels, and the necessity for parking spaces must be examined. It is important to maintain effective operation and maintenance of installed infrastructure. To maintain a seamless workflow, commercial players should be encouraged to participate rather than a primary government agency. Information communication between EV server-control centres will be greatly aided by a centralized system via charging station selection (CSS) servers.

2) Depleting potential energy sources: The second most frequently cited barrier to EV adoption is the alleged drawback of endurance [**18**]. The majority of individuals experience "range anxiety" because they are concerned that an EV won't have the scope to take them where they are required to go. Meanwhile organizations are expanding their products with improved efficiency and mileage. High power density of the battery is a crucial factor in making those capabilities possible. The cost of the goods will increase as a result of this. Especially nowadays, EVs have enough range to easily cover the daily commute of their users. Things like battery switching and improvements to the power grid for charging could assist.

3) Built-in power and a transit system: Concerning the electrical grid, electric vehicles confront special problems. The source of the recharging concern is the vehicle's flux in the power-consuming site, where power is pumped into the batteries out of the grid. Only the distribution grid is used to convey electric energy, which restricts the amount of energy that may flow through the conveyance network's transmission system. Another difficulty is determining if the power supply at the infrastructure of charging stations is regulated in real time with the flow of on-the-road automobiles. Grids that employ adaptable control characteristics may be capable of handling this workload well. Therefore, a data transmission of certain sort is required to serve as the open disclosure link between the transportation networks and the power grid, regardless of its type. As a result, the future electricity distribution grid will interact with the communication network, power infrastructure, and high quantity of EV charging load.

4) Smart grid requirement: In order to improve the power grid, additional renewable energy must be generated. However, the Indian electricity sector has not yet been able to fully match the demand. Employing the resources that are available wisely is another option. Installing new generation plants is a systematic procedure that will take some time in order to boost the supply. Now we're destined to need smart grids [**19**]. With the adoption of smart grids, communication between the utility provider and its consumers is facilitated by the ability to share information and electricity. The grid will become more effective, dependable, secure, and ecologically friendly thanks to its growing pattern of connections, controls, and new automation technologies and tools. The integration of various technologies into the smart grid makes the grid autonomous and effective [**20, 21**].

A lot of focus has been placed on Smart meters, as the accompanying picture illustrates. Smart meters are tools that help customers make decisions. Customers may decide when to turn their energy on or off and can change their consumption habits, for instance during peak times (as depicted in Figure 7.4).

5) Monetary concerns: Presently, the expenditure of the batteries placed in the cars makes acquiring an electric car more cost-effective than purchasing a combustible vehicle. Companies are collaborating with the production process for the assembly of electric car batteries to reduce prices and boost overall efficiency with the goal of reaching cost parity by 2025, if not

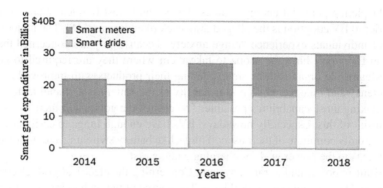

FIGURE 7.4 AI is the new electricity (Source: International Energy Agency. Mahendra, Ravi. 2019. Smart Energy International. May 11, 2019)

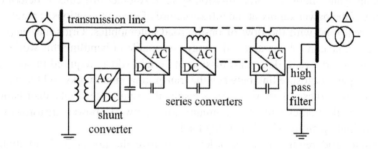

FIGURE 7.5 Block diagram of DPFC (Source: [36])

earlier. Having said that, federal subsidies are actively providing support in bringing expenditures to parity.

7.5 PROPOSED TECHNOLOGY

A few alluring benefits of AI application include: 1) EV cost cutting through ideal rechargeable batteries design and production, 2) precise scope estimation to reduce EV consumer range anxiety by predicting future road surfaces, 3) improved EV energy consumption when compared with traditional controls using AI safeguards for the EV assistant systems [22]), 4) a potential for increased road safety and optimal vehicular flow through connection and self-driving, and 5) lowered costs of EVs[23].

To ensure the benefits of AI technology, a system using artificial neural networks has been designed as a distributed power low controller (DPFC). DPFC is part of the flexible AC transmission system (FACTS) but is derived from the unified power flow controller (UPFC). The UPFC with an eliminated common DC link can be used for the DPFC [35]. In a UPFC, the exchange of active power takes place through the common DC link, but in a DPFC, this process occurs through the transmission lines at the third harmonic frequency.

TABLE 7.1

Various studies with their strategies (Source: [4])

Cite	Target	Strategy for Fixing Issues
[24]	Stability in regards to demand and supply while cutting expenses	Automated learning
[25]	Refuelling and discharging EVs to cut expenses and CO_2 emissions	Computer programming and PSO optimization
[26]	The projection for the generation of renewable energy is influenced by EV charging and discharge	Computational mathematics
[27]	A Q-learning technique that helps in monitoring commodity prices in a daily market and provides lucrative V2G services	Computational mathematics
[28]	Evaluation of V2G services using gameplay mechanics	Gamification and auctions
[29]	Coalitions amongst EVs to participate in the electricity markets and boost earnings	Gamification and auctions
[30]	PHEV collectives selling energy to the grid: a non-cooperative game to address the issue	Game theory and auctions
[31]	FREEDM system-based distributed grid intelligence for DER management	AI
[32]	ANN for decentralized grid utilization	AI
[33]	Distributed intelligence based on consensus for improving SG control	AI
[34]	Using Genetic Algorithm to optimize distributed generation operations	AI
Proposed	Increased system reliability by reducing large sized series 3-ϕ converters	AI

The concept of distributed FACTS (D-FACTS) is employed in the DPFC to eliminate the use of large size 3-ϕ series converter and use multiple small size 1-ϕ converters. The system reliability is increased by the use of a large number of series converters. Between the phases, there is no high voltage isolation, as the 1-ϕ converters are floating with respect to the ground, reducing the cost of the DPFC system when compared to that of the UPFC. Every converter inside the DPFC is free and has its own DC capacitor to give the required DC voltage. In addition, DPFC needs two Y-Δ transformers on either side of the transmission line and a shunt connected high-pass filter at the other end. The DPFC has a similar control capacity to the UPFC, and consecutive association between the shunt and series converters permits the trading of active power between converters. In a DPFC, the independence of the active power at various frequencies gives the likelihood that a converter without a power source can likewise produce active power at one recurrence and ingest this power from different frequencies.

A shunt converter simulation was carried out using an ANN controller, as shown in Figure 7.6.

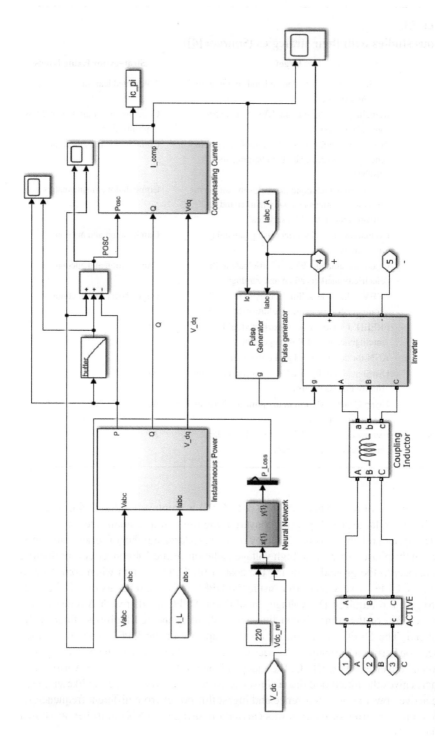

FIGURE 7.6 Shunt converter with ANN controller

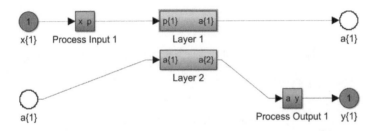

FIGURE 7.7 ANN controller

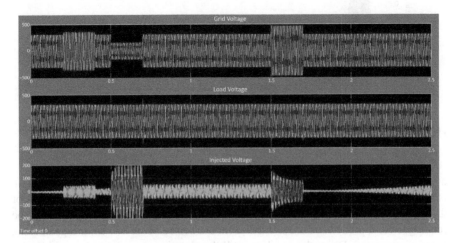

FIGURE 7.8 ANN controller load voltage, grid voltage, and injected voltage

It takes the actual and reference DC voltage to generate the power which is used to generate the gate pulses to the shunt converter. The ANN controller consists of layers of tuners for tuning purposes. The ANN controller is considered as the best as it has layers of tuners for fine tuning the signals.

Figure 7.7 represents the ANN controller used for the DPFC method. The ANN method gives the best results compared to everything, as this is used for high-level projects. Figure 7.8 shows the ANN controller voltages.

The total harmonic distortion (THD) plays an important role when it comes to the comparison of controllers [37]. Figure 7.9 represents the load voltages THD in fast Fourier transformation (FFT) analysis of the MATLAB/Simulink. The THD is found to be less (5.32%) for an ANN controller than for the other controllers. The grid voltages are considered important when it comes to calculation of harmonic distortions in the system.

7.6 OPERATION AND WORKING

To make recharging quicker and more productive, electric vehicles primarily require AI intelligence [38]. Fully electric and conventional ICEVs are not the same. There is

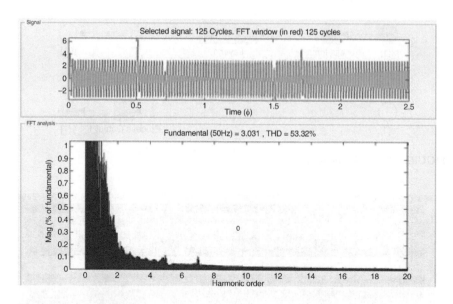

FIGURE 7.9   ANN controller grid side voltage

a lot more to it than just the fact that they are propelled uniquely. The key distinction is how quickly you might recharge them. By capping off an ICEV with gasoline or diesel, refuelling can be completed relatively rapidly. With an EV, which must be recharged for a predetermined number of hours before being driven, things become more complicated. Depending on how well the charging process works and how far the journey will take you, you will need to charge an EV for a certain amount of time. Businesses have invested a significant amount of effort into increasing the battery performance and creating larger batteries to make EVs efficient and practical. This has resulted in a major enhancement in the speed at which longer-distance cars may be charged.

Due to the chemical limitations of batteries, the expense, and the practical constraints of EV charging, there are a few major drawbacks regarding what can be accomplished moving ahead, even when considering the course of the upcoming 30 years. The administration of the recharging business itself, as well as the construction and operation of charging points, are examples of the physical constraints. The key requirement for AI in EV charging is to review the process, which represents one of the most significant obstacles to overcome in the field. While the EV is recharging, a procedure referred to as intelligent billing streamlines and optimizes choices. Amongst the goals of effective charging are the following: 1) Drastically reduce power demands at the charging location, 2) prevent excessive energy expenditures for both drivers and charging station providers, and 3) ensure punctual departures and an adequate level of charge for each vehicle.

The initial charging systems operated a few outlets per site or had very modest refining needs since they were constructed on very basic local embedded devices. For the majority of cases, this strategy is no longer practical because these devices

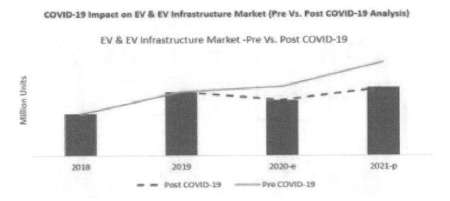

FIGURE 7.10   Impact on EV and EV infrastructure (Source: www.marketsandmarkets.com/ Market-Reports/covid-19-impact-on-electric-vehicle-market-181970499.html)

are often costly to install, prone to data shortage (failure to provide the most recent grid, vehicle, and price values), confined in their performance goals, or incompatible with smart charging driven by AI [39]). The way people see smart charging for electric vehicles has been completely redefined by several businesses. Designers were able to see through the restricted functionality that microcontrollers provided. Instead, a variety of fundamental requirements for fleet managers, energy suppliers, and producers of charging stations can be established:

Durability: The proposed AI-powered smart charging is reliable. The intention is to prevent personnel from configuring incorrectly and to completely rule out technical glitches on both the chargers and vehicle sides.

Extensibility: It is anticipated that recharging will become widely used and that the clients will want a battery charging application programming interface that is simple to scale.

Versatility: Mobility and smart recharging are related terms. The majority of smart power use cases call for dynamic decision-making, several improvement goals, and the gathering of information from diverse sources. The clients frequently wish to lower their energy bills, keep their departures on schedule, and lower the overall peak demand. Additionally, they would like the entire process to be computerized. A great deal of adaptability is needed for this.

For a long time, AI has been used in smartphones to help increase battery life, speed up charging, and maintain a higher charge. AI can, for example, monitor the battery's temperature to reduce the chance of it overheating. AI also improves the performance of electric vehicles which in turn improves the sales and demands for EVs. Presently, the power quality network should be more astute in providing a maintenance free and inexpensive power supply in balancing the continuously growing consumer interest in EV.

Smart grids offer more autonomous preventative measures and improved performance because of their innovative unique item alternatives. It could be possible to assume authority using visual and user experience technology. The V2G assistant aid

EV system configuration

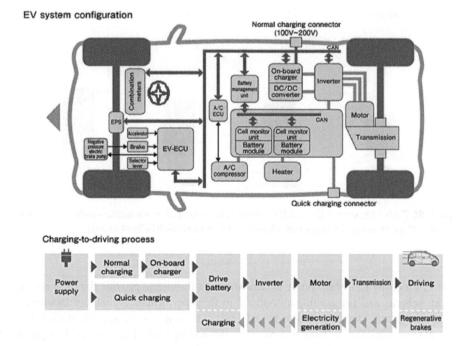

FIGURE 7.11 EV system configuration (Source: www.st.com/en/applications/electro-mobility/automotive-battery-management-system-bms.html)

framework uses administration-organized technology, ideal control, and composed control hypotheses to guide the communicated management structure.

The specific application of management-focused innovation depends on the power grid's consistent operation and EV buyers' actual charging needs. Most experts in the field concur that the technique may hasten all phases of the continuous battery improvement, from player science to estimation and frame assurance to superior materials and invention setups.

Future EV specialists will be completely able to access most of the electrical components operating process (as shown in Figure 7.11) using AI techniques. A sophisticated computer or computer automaton's capacity to carry out tasks that are typically performed by sentient individuals. The process of AI in creating techniques with mental repetitions that are similar to those of humans, such as the ability to process, take initiative, summarize, or learn from past knowledge. Artificial intelligence (AI) in vehicle production helps automakers save assembly costs while also supplying a safer and more productive manufacturing line environment [40]. Utilizing technological developments like computer vision, it is simple to spot product flaws.

7.7 RESULTS AND COMPARATIVE ANALYSIS

Figure 7.12 depicts the sales of regional registered EVs in India. Positive changes have also been made in the country's efforts to expand the availability of charging stations; states with aspirational targets for installing public charging points to

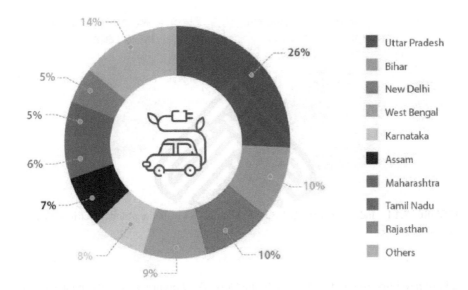

FIGURE 7.12 Pie chart (Source: www.india-briefing.com/news/electric-vehicle-industry-in-india-why-foreign-investors-should-pay-attention-21872.html/)

EV Sales for Fiscal Year Ending March 2020			
Segment	FY 2019	FY 2020	Change in percentage
Cars	3,600	3,400	-5
Two wheelers	126,000	152,000	21
Buses	400	600	50

FIGURE 7.13 EV sales for two-wheelers in FY 2020 (Source: Quartz/SMEV)

support the nation's adoption of electric automobiles include Andhra Pradesh, Uttar Pradesh, Bihar, and Telangana. The major reasons why certain states are operating substantially better than some others include good local policies, enhanced transportation, investor-friendly governmental policies, and professional development through quicker access to regulators, supply chain links, and the availability of sufficient land.

Karnataka became the first Indian state to pass a comprehensive EV law, and it has ever since developed into a centre for the nation's EV sector in terms of both manufacturing and R&D for EVs and associated necessary equipment. The supply chain, available land, closeness to ports, and aggressive investment help provided by government websites like Guidance Tamil Nadu are all contributing to Tamil Nadu's outstanding rate of growth. Although the EV industry is expanding, it still has to make substantial advancements if it is to fulfil the government's 2030 goal. In

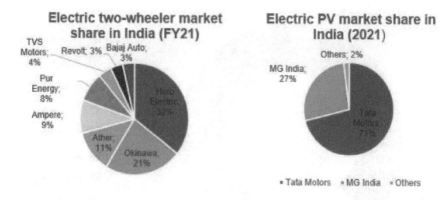

FIGURE 7.14 Indian market EV sales (Source: www.ibef.org/blogs/electric-vehicles-market-in-india)

addition to impeding industry growth, the COVID-19 outbreak also caused a general decline in demand.

However, in several markets, consumer confidence has remained upbeat. India had a 21% growth in EV sales for two-wheelers in FY 2020, in Figure 7.13. Sales of EV buses rose by 50% within the same time period. On the other hand, the market for EVs continued to decline, falling by 5%. Regarding overall EV sales, they seem to be gradually increasing after experiencing a setback in 2020. Sales of electric vehicles (EVs) totalled 15,910 in India in January 2021, with the state of Uttar Pradesh selling the most of them, followed by Bihar and Delhi.

As close to 0.32 million vehicles were sold in 2021, the Indian EV market (as shown in Figure 7.14) is expanding quickly. Sales were up 168% year over year in this case. The Paris Agreement, which aims to cut carbon emissions and improve air quality, is a major driver of the electric vehicle industry expansion in India. Three companies—Hero Electric, Okinawa, and Ather Energy—dominate the Indian electric vehicle industry. Collectively, they control 64% of the market. The different state policies in India (Table 7.2) for electric vehicle are as follows:

7.8 CONCLUSION

The nicest thing about these electric vehicles is that they support the AI function, which creates a hands-free driving experience. They also support navigation systems and behaviour monitoring. The AI system in electric vehicles offers real-time driving monitoring and energy use optimization, which will allow for the additional driving range and provide riders fewer panic attacks due to low battery. In the future, a wider target audience will be drawn to the smart grid access points and charging stations. The EV models make use of V2G, which provides EV owners with an extra source of revenue. Artificial intelligence is playing a huge role in the EV industry, with applications including autonomous driving, client behaviour monitoring, and creative route frameworks. It may very well be used for health purposes, such as supporting

TABLE 7.2

Policies for EVs State-wise (Source: www.india-briefing.com/news/electric-vehicle-industry-in-india-why-foreign-investors-should-pay-attention-21872.html/)

	State Policies for EVs
Assam state, 2021	• Released September 2021, for a period of five years. • Obtain a 25% penetration rate for EVs in Assam's overall vehicle registrations. • Bolster the five-year deployment of 200,000 EVs. • This target's segment breakdown is as follows: • 100,000 two-wheeled EVs • 75,000 three-wheeled EVs
Gujarat state, 2021	• 1,10,000 two-wheeled EVs • 70,000 three-wheel EVs • 20,000 four-wheel EVs • Based on battery capacity, incentives for EVs will be given up to INR 10,000 (US$134.40)/kWh. • There will be no registration fees for any EVs. • Policy incentives to improve the state's infrastructure for charging.
Rajasthan, 2021	• State Goods and Services Tax (SGST) refunds are available for all electric vehicles purchased before March 2022. • Additional financial assistance for the purchase of electric two- and three-wheelers.
Meghalaya, 2021	• Launched in March 2021, valid for five years following notification. • Encourage the purchase of 20,000 EVs while the policy is being implemented. • The payment of registration fees and road tax shall not be required for any kind of EV purchased during the term of the policy. • By encouraging private investment, charging infrastructure is improved. • Encourage battery recycling and reuse.
West Bengal, 2021	• During the period when the policy is being implemented, the state should have one million EVs across all market segments. • In the next five years, it is planned to build 100,000 public and semi-public charging stations. • Reach an eight-point EV/public charge point ratio. • Old batteries can be recycled, reused, and disposed of in an environmentally friendly way. • Construction of an "EV Accelerator Cell." • Through DISCOMs, make public EV charging infrastructure possible. • By 2024, the goal is one million EVs.
Andhra Pradesh	• According to the government, it also plans to stop registering gasoline and diesel vehicles in the upcoming capital city of Amaravati.
Maharashtra	• 10% of all new vehicle enrolments by 2025 ought to be electric cars. • Achieve a 25% electrification rate for mass transit and last-mile delivery vehicles by 2025 in South Mumbai, Maharashtra, Nashik, and Aurangabad. • Create charging stations and highways that connect them throughout the state. Rewards for installing charging stations.

(Continued)

TABLE 7.2 (Continued)

Policies for EVs State-wise (Source: www.india-briefing.com/news/electric-vehicle-industry-in-india-why-foreign-investors-should-pay-attention-21872. html/

	State Policies for EVs
Karnataka	• 1,000 electric buses will be added to local public transportation bus fleets. • In Bengaluru, trying to install 112 EV charging stations. • Focus on developing a secondary market for batteries and venture capital funds for emerging e-mobility companies. • Incentives for EV manufacturing businesses, like interest-free loans based on net SGST.
Odisha	• Dated August 2021; released. Expires after five years. • By 2025, all new vehicle registrations in the state must include 20% electric vehicles. • Road tax and registration fees during the term of the policy are waived. • Incentives for the production of batteries and other EV parts.
Himachal Pradesh	• By 2030, all vehicles will be electric. • The blueprint contains a provision for charging stations in structural applications and encourages the development of specialized charging networks.
Tamil Nadu	• Within ten years, replace all auto-rickshaws in six important cities with EVs. • Promote the start-up of electric vehicle companies, create hubs for venture capital and business incubation services. • Manufacturing facilities for EV-related and charging infrastructure will be exempt from electricity tax at 100% until 2025.
Madhya Pradesh	• New ICEVs won't be registered in some cities. • Ensure a secure, affordable, and easily accessible charging infrastructure to promote faster adoption. • Incentives for shared e-rickshaws and electric auto-rickshaws include no permit fees, a five-year exemption or reimbursement from road taxes and vehicle registration fees, and a five-year 100% parking fee waiver at any parking facility operated by a municipal corporation.
Bihar	• Rickshaw power generation should be prioritized. By 2022, all paddle rickshaws are to be replaced with electric rickshaws. • Promotion of e-rickshaw production. • Establish charging stations at commercial and residential locations as well as fast-charging stations on state and national highways spaced 50 km apart.

vision equipment, assessing driver behaviour, and ensuring vehicle security. Due to a number of variables, including a strong economy, social transformation, fast urbanization, good segment earnings, and so forth, India's financial growth is evidently headed for positive development. The ease with which work records may now be maintained in India is evidence of the nation's increased capacity to establish and support organizations.

This chapter describes the ANN concepts in order to show how AI techniques can be implemented to achieve a coordinated operation in between EVCSs and distributed power generation in the grid. The scheduling of energy sales and several auxiliary services can also be performed by developing suitable AI algorithm.

REFERENCES

1. L. Luo, Z. Wu, W. Gu, H. Huang, S. Gao, and J. Han, "Coordinated allocation of distributed generation resources and electric vehicle charging stations in distribution systems with vehicle-to-grid interaction," *Energy*, 29-Nov-2019. [Online]. Available: www.sciencedirect.com/science/article/pii/S0360544219323266.
2. C.-T. Ma, "System planning of grid-connected electric vehicle charging stations and key technologies: A review," *MDPI*, 04-Nov-2019. [Online]. Available: www.mdpi.com/1996-1073/12/21/4201.
3. N. Etherden, V. Vyatkin, and M. H. Bollen, "Virtual power plant for grid services using IEC 61850," *IEEE Transactions on Industrial Informatics*, vol. 12, no. 1, pp. 437–447, 2016.
4. E. S. Rigas, S. D. Ramchurn, and N. Bassiliades, "Managing electric vehicles in the smart grid using artificial intelligence: A survey," *IEEE Transactions on Intelligent Transportation Systems*, vol. 16, no. 4, pp. 1619–1635, 2015.
5. S. Mahdavi, R. Hemmati, and M. A. Jirdehi, "Two-level planning for coordination of energy storage systems and wind-solar-diesel units in active distribution networks," *Energy*, 23-Mar-2018. [Online]. Available: www.sciencedirect.com/science/article/pii/S0360544218305322.
6. S. Davidov and M. Pantoš, "Optimization model for charging infrastructure planning with electric power system reliability check," *Energy*, 26-Oct-2018. [Online]. Available: www.sciencedirect.com/science/article/pii/S0360544218321467.
7. M. H. Amini, M. P. Moghaddam, and O. Karabasoglu, "Simultaneous allocation of electric vehicles' parking lots and distributed renewable resources in smart power distribution networks," *Sustainable Cities and Society*, 19-Oct-2016. [Online]. Available: www.sciencedirect.com/science/article/pii/S2210670716304966.
8. H. Zhang, S. Moura, Z. Hu, W. Qi, and Y. Song, "Joint PEV charging station and distributed PV generation planning," in *2017 IEEE Power & Energy Society General Meeting*, 2017.
9. H. Zhang, S. J. Moura, Z. Hu, W. Qi, and Y. Song, "Joint PEV charging network and distributed PV generation planning based on accelerated generalized benders decomposition," *IEEE Transactions on Transportation Electrification*, vol. 4, no. 3, pp. 789–803, 2018.
10. S. M. Kandil, H. E. Z. Farag, M. F. Shaaban, and M. Z. El-Sharafy, "A combined resource allocation framework for PEVS charging stations, renewable energy resources and distributed energy storage systems," *Energy*, 03-Nov-2017. [Online]. Available: www.sciencedirect.com/science/article/pii/S0360544217318546.
11. E. Sortomme and M. A. El-Sharkawi, "Optimal charging strategies for unidirectional vehicle-to-grid," *IEEE Transactions on Smart Grid*, vol. 2, no. 1, pp. 131–138, 2011.
12. S. Gao, K. T. Chau, C. Liu, D. Wu, and C. C. Chan, "Integrated energy management of plug-in electric vehicles in power grid with renewables," *IEEE Transactions on Vehicular Technology*, vol. 63, no. 7, pp. 3019–3027, 2014.

13. H. Morais, T. Sousa, Z. Vale, and P. Faria, "Evaluation of the electric vehicle impact in the power demand curve in a smart grid environment," *Energy Conversion and Management*, 01-Apr-2014. [Online]. Available: www.sciencedirect.com/science/article/pii/S0196890414002246.

14. L. Jian, X. Zhu, Z. Shao, S. Niu, and C. C. Chan, "A scenario of vehicle-to-grid implementation and its double-layer optimal charging strategy for minimizing load variance within regional smart grids," *Energy Conversion and Management*, 08-Dec-2013. [Online]. Available: www.sciencedirect.com/science/article/pii/S0196890413007280.

15. E. Sortomme and M. A. El-Sharkawi, "Optimal scheduling of vehicle-to-grid energy and ancillary services," *IEEE Transactions on Smart Grid*, vol. 3, no. 1, pp. 351–359, 2012.

16. Available: https://www.smart-energy.com/industry-sectors/cybersecurity/how-hackers-target-smart-meters-to-attack-the-grid/

17. H. Karimipour, A. Dehghantanha, R. M. Parizi, K.-K. R. Choo, and H. Leung, "A deep and scalable unsupervised machine learning system for cyber-attack detection in large-scale smart grids," *IEEE Access*, vol. 7, pp. 80778–80788, 2019.

18. S. S. Ali and B. J. Choi, "State-of-the-art artificial intelligence techniques for distributed smart grids: A review," *MDPI*, 22-Jun-2020. [Online]. Available: www.mdpi.com/2079-9292/9/6/1030.

19. R. Couillet, S. M. Perlaza, H. Tembine, and M. Debbah, "Electrical vehicles in the smart grid: A mean field game analysis," *IEEE Journal on Selected Areas in Communications*, vol. 30, no. 6, pp. 1086–1096, 2012.

20. J. Li, Y. Zhao, C. Sun, X. Bao, Q. Zhao, and H. Zhou, "A survey of development and application of artificial intelligence in smart grid," *IOP Conference Series: Earth and Environmental Science*, vol. 186, p. 012066, 2018.

21. R. Yaqub, S. Ahmad, H. Ali, and A. ul Asar, "AI and blockchain integrated billing architecture for charging the roaming electric vehicles," *MDPI*, 15-Nov-2020. [Online]. Available: www.mdpi.com/2624-831X/1/2/22.

22. O. A. Omitaomu and H. Niu, "Artificial intelligence techniques in smart grid: A survey," *MDPI*, 22-Apr-2021. [Online]. Available: www.mdpi.com/2624-6511/4/2/29.

23. M. Ahmed, Y. Zheng, A. Amine, H. Fathiannasab, and Z. Chen, "The role of artificial intelligence in the mass adoption of electric vehicles," *Joule*, 11-Aug-2021. [Online]. Available: www.sciencedirect.com/science/article/pii/S2542435121003500.

24. S. Chatzivasileiadis, M. D. Galus, Y. Reckinger, and G. Andersson, "Q-learning for optimal deployment strategies of frequency controllers using the aggregated storage of PHEV fleets," in *2011 IEEE Trondheim PowerTech*, 2011.

25. A. Y. Saber and G. K. Venayagamoorthy, "Plug-in vehicles and renewable energy sources for cost and emission reductions," *IEEE Transactions on Industrial Electronics*, vol. 58, no. 4, pp. 1229–1238, 2011.

26. M. Galus and G. Andersson, "Balancing renewable energy source with vehicle to grid services from a large fleet of plug-in hybrid electric vehicles controlled in a metropolitan area distribution network towards the power system of the future modelling/new tools for technical performance assessment," *Semantic Scholar*, 01-Jan-1970. [Online]. Available: www.semanticscholar.org/paper/Balancing-Renewable-Energy-Source-with-Vehicle-to-a-Galus-Andersson/741a6e0fde22b1bb8cfe5622319cfe072a683689.

27. L. Wehinger, G. Hug, M. Galus, and G. Andersson, "Assessing the effect of storage devices and a PHEV cluster on German spot prices by using model predictive and profit maximizing agents," *Semantic Scholar*, 01-Jan-1970. [Online]. Available: www.semanticscholar.org/paper/Assessing-the-Effect-of-Storage-Devices-and-a-PHEV-Wehinger-Hug/fb178727825b89bff14ef8d1d0d59c1080cb2234.

28. S. Kamboj, W. Kempton, and K. S. Decker, "[PDF] deploying power grid-in-tegrated electric vehicles as a multi-agent system: Semantic scholar," *Undefined*, 01-Jan-1970. [Online]. Available: www.semanticscholar.org/paper/Deploying-power-grid-integrated-electric-vehicles-a-Kamboj-Kempton/f1ef7b1f8be0b20c52f7c18682250d5d9ce148cf.

29. G. de Ramos, J. C. Rial, and A. L. C. Bazzan, "Self-adapting coalition formation among electric vehicles in smart grids," in *2013 IEEE 7th International Conference on Self-Adaptive and Self-Organizing Systems*, 2013.

30. W. Saad, Zhu Han, H. V. Poor, and T. Basar, "A noncooperative game for double auc-tion-based energy trading between PHEVs and distribution grids," in *2011 IEEE International Conference on Smart Grid Communications (SmartGridComm)*, 2011.

31. F. Meng, R. Akella, M. L. Crow, and B. McMillin, "Distributed grid intelligence for future microgrid with renewable sources and storage," *North American Power Symposium 2010*, 2010.

32. M. H. Elkazaz, A. A. Hoballah, and A. M. Azmy, "Operation optimization of distrib-uted generation using artificial intelligent techniques," *Ain Shams Engineering Journal*, 18-Feb-2016. [Online]. Available: www.sciencedirect.com/science/article/pii/S2090447916000253.

33. K. Utkarsh, A. Trivedi, D. Srinivasan, and T. Reindl, "A consensus-based distributed computational intelligence technique for real-time optimal control in smart distribution grids," *IEEE Transactions on Emerging Topics in Computational Intelligence*, vol. 1, no. 1, pp. 51–60, 2017.

34. M. N. Q. Macedo, J. J. M. Galo, L. A. L. de Almeida, and A. C. de C. Lima, "Demand side management using artificial neural networks in a smart grid environment," *Renewable and Sustainable Energy Reviews*, 06-Sep-2014. [Online]. Available: www.science-direct.com/science/article/pii/S1364032114007114.

35. Z. Yuan, S. W. H. de Haan, J. B. Ferreira and D. Cvoric, "A FACTS Device: Distributed Power-Flow Controller (DPFC)," in *IEEE Transactions on Power Electronics*, vol. 25, no. 10, pp. 2564–2572, Oct. 2010, doi: 10.1109/TPEL.2010.2050494.

36. A. K. Bahamani, G. M. Sreerama Reddy, and V. Ganesh. "Power quality improvement in fourteen bus system using non conventional source based ANN controlled DPFC," *Indonesian Journal of Electrical Engineering and Computer Science,* vol. 4, no. 3, 2016.

37. M. N and R. H. R, "Comparative analysis of modulation techniques for elimination of CMV in multilevel inverters," in *2022 Fourth International Conference on Emerging Research in Electronics, Computer Science and Technology (ICERECT)*, 2022, pp. 1–11, doi: 10.1109/ICERECT56837.2022.10060103.

38. Dr. Varadaraj Aravamudhan, Dr. Ananth Sengodan, and Dr. Prasanna Sai Mohanraj, "Impact of artificial intelligence on EV industry and perceptual mapping in India," vol. 17, no. 7, pp. 1360–1373, Jul-2022, doi: 10.5281/zenodo.6939189.

39. M. D. Lytras and K. T. Chui, "The recent development of artificial intelligence for smart and sustainable energy systems and applications," *Energies*, vol. 12, no. 16, p. 3108, 2019.

40. B. Makala and T. Bakovic, "Artificial intelligence in the power sector," *Open Knowledge Repository*, 01-Apr-2020. [Online]. Available: https://openknowledge.worldbank.org/handle/10986/34303.

8 Model Predictive Control of Grid-Connected Wind Energy Conversion System Using VSC-Based Shunt Controllers

Apoorva Srivastava and R.S. Bajpai

This chapter presents a novel approach to controlling the voltage of a feeder using model predictive control (MPC). This approach generates an equivalent phase angle delta, thereby helping in maintaining the energy balance at the electricity network nodes even when wind and load changes occur. The efficiency of this strategy was assessed using both MATLAB/Simulink simulation studies and a hardware proto-type executed in the one-phase delivery network using the dSPACE platform. The results demonstrate that the proposed approach is beneficial for optimizing energy in hybrid renewable power systems.

8.1 INTRODUCTION TO DISTRIBUTED ENERGY RESOURCES

Power electronics controllers play an important role in extracting maximum power from variable-speed wind turbines and providing various option to integrate wind systems into the grid in order to achieve higher efficiency and better power system performance. Voltage source converter (VSC)-based shunt controllers are often employed in the distributed generation for load compensation. Incorporating battery storage into shunt or series VSCs can help improve power quality by mitigating issues such as short interruptions, voltage sags and swells, and harmonic distortions in the grid. When the device is powered by a wind energy conversion system, it can enhance the voltage quality and power flow of the distribution system. This will be a value-added benefit to the controller in terms of regulating the grid's terminal voltage and compensating the grid voltage to balance sinusoid, and excess wind energy will be returned to other feeds, reducing power cuts and the possibility of major blackouts.

MPC has been shown to be an effective means of operating power converters that are connected to the grid, even when the system has non-linearities and restrictions. MPC has recently been used to improve the efficacy and seamless operation of AC microgrids by regulating grid-connected power converters. Under this control scheme, a subsystem model is developed to forecast the behavior of control variables

DOI: 10.1201/9781003311829-8

for a defined period in the future. These predictions are assessed using the cost function in order to implement the optimum switching states. The single-line schematic of the energy delivery system using a VSC is shown in Figure 8.1. The wind turbine generator system provides input energy to controller. The wind turbine, a permanent magnet synchronous generator (PMSG), an unregulated diode rectifier circuits, a DC–DC conversion device, a model forecasting control system that tracks optimum power points, and a VSC are all included in the control scheme, as shown in Figure 8.1. The employment of the PMSG alongside a varying-rate wind turbine is chosen since it provides the maximum power point tracker (MPPT) with a wide range of power control. By replacing field excitation and transmission gears, direct drive PMSGs have lower maintenance costs than doubly fed induction generator, resulting in improved energy efficiency and longevity.

8.2 VOLTAGE AND POWER FLOW CONTROL DEVICES

Renewable energy sources such as wind turbines can improve distribution system stability by using their full capacity to satisfy the load demand. Additionally, improving voltage quality and load sharing by injecting wind power into the grid at different occasions may be economically viable, preventing load shedding and the threat of future blackouts.

Generally, the Distribution Static Compensator, a shunt-connected device, is a very attractive option for voltage sag minimization in weak grids. It is widely employed on the distribution side to eliminate current harmonics, compensate for reactive power, and reduce voltage disturbances that cause flicker [1].

For voltage perturbation reduction and load bus voltage regulation, VSC-based operating systems are widely employed [2, 3]. VSC-based shunt controllers can be operated in current control mode [4] and voltage-control mode [5–7]. They can minimize harmonic distortion, unbalanced load components, and adjust the grid voltage to a balanced sinusoid in voltage-control mode, regardless of source or load disturbances.

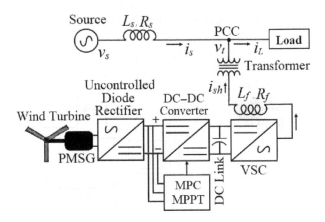

FIGURE 8.1   Single-line schematic of a control scheme

8.3 MPPT CONTROL STRATEGY

Because of the uncertainty of wind and its influence on grid integration, for wind energy conversion systems, it is essential to test the efficacy of the control algorithm under laboratory conditions. The design of control systems for wind turbines with a rating of 10 kW or higher has been studied in the proposed research [8].

In the literature [9–13], it is suggested to conduct a study on simulators for wind turbines. A wind turbine emulator using a DC motor was employed using a two-mass oscillatory system that closely resembles the features of genuine wind turbines.

Figure 8.2 depicts a model forecasting approach that uses the standard incremental conductance (INC) methodology to generate the optimal amount of electricity from a wind turbine-controlled energy system. The MPC MPPT technique is proposed for grid-connected systems in this chapter. To evaluate the effectiveness of the proposed MPC tracking system, a simulation experiment is performed utilizing varied wind characteristics, as illustrated in the following case studies.

Case 8.3.1: As demonstrated in Figure 8.3, the MPPT controller first stabilized at its optimal MPP. A step rise in speed of from 10 m/s to 12 m/s is introduced to the wind turbine system at t = 10 s. As demonstrated in Figure 8.3, the proposed MPC MPPT has a faster detection performance and negligible disturbances in the equilibrium state when compared to the traditional INC MPPT. Figure 8.3 depicts the proposed model forecasting approach. It is capable of producing more energy than the traditional INC MPPT approach.

Case 8.3.2: The recommended MPPT scheme is evaluated in a wind turbine system with multi-step wind speed fluctuations. As illustrated in Figure 8.4, the speed of the wind is previously set at 8 m/s, and subsequently raised from 8 m/s to 10 m/s and from 10 m/s to 12 m/s at t = 10 s and t = 12 s, respectively. Figure 8.5 shows that the proposed prediction approach for power tracking achieves a faster tracking response than the traditional INC MPPT despite rapidly fluctuating wind speeds.

Case 8.3.3: As illustrated in Figure 8.5, the system has been brought to a state of equilibrium at a wind speed of 12 m/s with the implementation of the suggested MPPT controller. Wind speed decreases from 12 to 10 m/s at $t = 10$ s and increases from 10 to 12 m/s at $t = 12$ s. According to the findings shown in Figure 8.5, the suggested prediction strategy can more rapidly follow the MPP compared to the

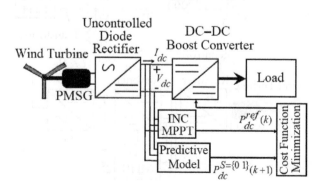

FIGURE 8.2   MPC MPPT strategy

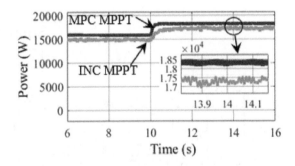

FIGURE 8.3   Illustration of power captured from INC and proposed MPC MPPT

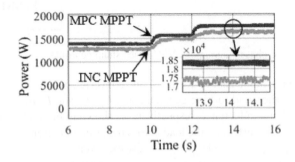

FIGURE 8.4   Illustration of power captured from INC and proposed MPC MPPT

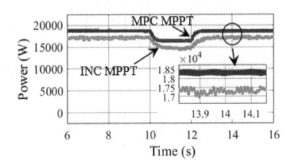

FIGURE 8.5   WT power (P_{dc}) with INC MPPT and proposed MPC MPPT

traditional INC method. Findings showed that using a typical INC technique alone does not provide as much accuracy as combining it with prediction.

8.4   FINITE SET CONTROL MPC STRATEGIES

As shown in Figure 8.6, a VSC -based shunt controller is developed to execute the suggested MPC approach for controlling voltage and power flow. The wind turbine

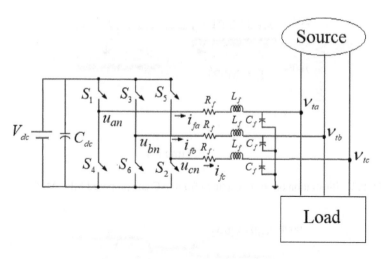

FIGURE 8.6   Three-phase diagram of a VSC-based shunt controller

generator provides input energy to the shunt converter, where V_{dc} is the rectified DC voltage from the wind turbine generator and C_{fa}, R_f and L_f are the filter capacitance, resistance, and inductance, respectively, as shown in Figure 8.6. The electrical parameters of the converter and the load across the point of common coupling (PCC) are shown by u_{an}, u_{bn}, u_{cn} and v_{ta}, v_{tb}, v_{tc}, respectively.

Kirchhoff's law may be used to analyze the circuit in Figure 8.6.

$$L_f \frac{d}{dt} \begin{bmatrix} i_{fa} \\ i_{fb} \\ i_{fc} \end{bmatrix} + R_f \begin{bmatrix} i_{fa} \\ i_{fb} \\ i_{fc} \end{bmatrix} = \begin{bmatrix} u_{an} \\ u_{bn} \\ u_{cn} \end{bmatrix} - \begin{bmatrix} v_{ta} \\ v_{tb} \\ v_{tc} \end{bmatrix} \tag{8.1}$$

Through the use of the Clarke transformation, Equation (8.1) can be converted to the α, β frame, which can be expressed as

$$L_f \frac{d}{dt} \begin{bmatrix} i_\alpha \\ i_\beta \end{bmatrix} + R_f \begin{bmatrix} i_\alpha \\ i_\beta \end{bmatrix} = \begin{bmatrix} u_\alpha \\ u_\beta \end{bmatrix} - \begin{bmatrix} v_\alpha \\ v_\beta \end{bmatrix} \tag{8.2}$$

The magnitudes of the output voltages from the VSC, as shown in Figure 8.6, may be described as state space vectors.

$$AS_a = \begin{bmatrix} 1, & \textit{if } S_1 \textit{ ON and } S_4 \textit{ OFF} \\ 0, & \textit{if } S_1 \textit{ OFF and } S_4 \textit{ ON} \end{bmatrix}$$

$$AS_b = \begin{bmatrix} 1, & \textit{if } S_2 \textit{ ON and } S_5 \textit{ OFF} \\ 0, & \textit{if } S_2 \textit{ OFF and } S_5 \textit{ ON} \end{bmatrix} \tag{8.3}$$

$$AS_c = \begin{bmatrix} 1, & \textit{if } S_3 \textit{ ON and } S_6 \textit{ OFF} \\ 0, & \textit{if } S_3 \textit{ OFF and } S_6 \textit{ ON} \end{bmatrix}$$

Equation (8.3) can be expressed in a generalized form as

$$AS = \frac{2}{3}\left(AS_a + aAS_b + a^2 AS_c\right)$$

(8.4)

Where $a = e^{j2\pi/3}$.

By representing the converter output voltages as state space vectors, we can formulate them as

$$\begin{bmatrix} u_{an} \\ u_{bn} \\ u_{cn} \end{bmatrix} = \frac{V_{dc}}{3} \begin{bmatrix} 2 & -1 & -1 \\ -1 & 2 & -1 \\ -1 & -1 & 2 \end{bmatrix} \begin{bmatrix} AS_a \\ AS_b \\ AS_c \end{bmatrix}$$

(8.5)

By applying the Clarke transformation, Equation (8.5) can be expressed in the α, β frame as

$$\begin{bmatrix} u_\alpha \\ u_\beta \end{bmatrix} = \frac{\sqrt{6}V_{dc}}{9} \begin{bmatrix} 1 & -\dfrac{1}{2} & -\dfrac{1}{2} \\ 0 & \dfrac{\sqrt{3}}{2} & -\dfrac{\sqrt{3}}{2} \end{bmatrix} \times \begin{bmatrix} 2AS_a & -AS_b & -AS_c \\ -AS_a & 2AS_b & -AS_c \\ -AS_a & -AS_b & 2AS_c \end{bmatrix}$$

(8.6)

As demonstrated, the current delivered to the grid through a VSC-based shunt controller can be discretized as

$$\frac{di_f}{dt} = \frac{i_f(k) - i_f(k-1)}{T_s}$$

(8.7)

T_s stands for sampling time.

By converting Equation (8.2) to its discrete form, we can express it as

$$\frac{L_f}{T_s}\begin{bmatrix} i_\alpha(k) - i_\alpha(k-1) \\ i_\beta(k) - i_\beta(k-1) \end{bmatrix} + R_f \begin{bmatrix} i_\alpha(k) \\ i_\beta(k) \end{bmatrix} = \begin{bmatrix} u_\alpha(k) \\ u_\beta(k) \end{bmatrix} - \begin{bmatrix} v_{t\alpha}(k) \\ v_{t\beta}(k) \end{bmatrix}$$

(8.8)

Consequently, predicted currents may be calculated as

$$\begin{bmatrix} i_\alpha(k) \\ i_\beta(k) \end{bmatrix} = \frac{L_f}{L_f + RT_s}\begin{bmatrix} i_\alpha(k-1) \\ i_\beta(k-1) \end{bmatrix} + \frac{T_s}{L_f + RT_s}\begin{bmatrix} u_\alpha(k) - v_{t\alpha}(k) \\ u_\beta(k) - v_{t\beta}(k) \end{bmatrix}$$

(8.9)

The following is a formulation for forecasting current shifting

$$\begin{bmatrix} i_\alpha(k+1) \\ i_\beta(k+1) \end{bmatrix} = \left(\frac{L_f}{L_f + RT_s}\right)\begin{bmatrix} i_\alpha(k) \\ i_\beta(k) \end{bmatrix} + \frac{T_s}{L_f + RT_s}\begin{bmatrix} u_\alpha(k+1) - v_{t\alpha}(k+1) \\ u_\beta(k+1) - v_{t\beta}(k+1) \end{bmatrix}$$

(8.10)

where $i_\alpha(k), i_\beta(k), u_\alpha(k), u_\beta(k), v_{t\alpha}(k)$, and $v_{t\beta}(k)$ are the $\alpha\beta$ components of output currents, voltages, and grid voltages, respectively.

Applying Equation (8.10) allows for the evaluation of the estimated grid voltage vector

$$\begin{bmatrix} v_{t\alpha}(k) \\ v_{t\beta}(k) \end{bmatrix} = \frac{L_f}{T_s}\begin{bmatrix} i_\alpha(k-1) \\ i_\beta(k-1) \end{bmatrix} - \frac{L_f + RT_s}{T_s}\begin{bmatrix} i_\alpha(k) \\ i_\beta(k) \end{bmatrix} + \begin{bmatrix} u_\alpha(k) \\ u_\beta(k) \end{bmatrix} \tag{8.11}$$

If the assumption is made that the grid voltage doesn't change instantly throughout a sampling time, then a one-step forward predicted voltage can be calculated.

$$\begin{bmatrix} v_{t\alpha}(k+1) \\ v_{t\beta}(k+1) \end{bmatrix} = \frac{L_f}{T_s}\begin{bmatrix} i_\alpha(k) \\ i_\beta(k) \end{bmatrix} - \frac{L_f + RT_s}{T_s}\begin{bmatrix} i_\alpha(k+1) \\ i_\beta(k+1) \end{bmatrix} + \begin{bmatrix} u_\alpha(k+1) \\ u_\beta(k+1) \end{bmatrix} \tag{8.12}$$

A cost function (g) is developed to manage load voltage and enable adequate distribution of power from VSC supported by the windmill.

$$g = \left(v_{t\alpha}^*(k) - v_{t\alpha}(k+1)\right)^2 + \left(v_{t\beta}^*(k) - v_{t\beta}(k+1)\right)^2 \tag{8.13}$$

The recommended control mechanism and appropriate switching state decision-making process for optimizing the recommended cost function are depicted in Figure 8.7. Figure 8.8 shows a flowchart of the MPC algorithmic structure.

8.5 POWER FLOW CONTROL STRATEGY

The following equations explain the allocation of electricity among sources, loads, and wind turbines

$$P = \frac{V_s V_t \, \mathrm{Sin}\, \delta}{X} \tag{8.14}$$

$$Q = \frac{V_s^2 - V_s V_t \, \mathrm{Cos}\, \delta}{X} \tag{8.15}$$

In this case, δ is the phase difference between voltages V_s and V_t, and X represents the distribution line's reactance. Voltages V_s and V_t are assessed in the preceding Equations (8.14) and (8.15) as the root mean square values (RMS) of the instantly voltage magnitudes v_s and v_t.

The effect of altering phase angle delta on real power (P) and reactive power (Q) may be monitored properly using equations (8.14) and (8.15).

$$\frac{\partial P}{\partial \delta} = \frac{V_s V_t}{X} \, \mathrm{Cos}\, \delta \tag{8.16}$$

$$\frac{\partial P}{\partial V_s} = \frac{V_t}{X} \, \mathrm{Sin}\, \delta \tag{8.17}$$

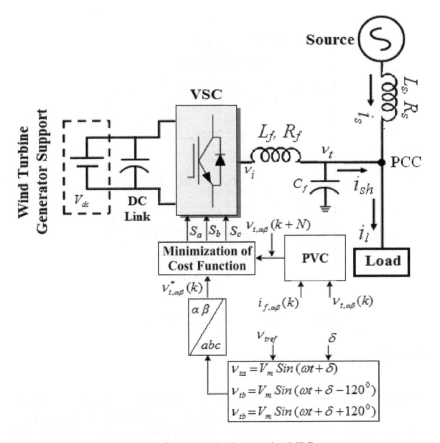

FIGURE 8.7   Voltage and power flow control scheme using MPC

$$\frac{\partial Q}{\partial '} = \frac{V_s\, V_t}{X}\, \mathrm{Sin}\,\delta \tag{8.18}$$

$$\frac{\partial Q}{\partial V_s} = \frac{2V_s - V_t\, \mathrm{Cos}\,\delta}{X} \tag{8.19}$$

In actual practice, phase angle δ is typically minimal; therefore, $\partial P\,/\,\partial\delta$ is closely approaching $V_s\, V_t\,/\,X$ and $\partial P\,/\,\partial V_s$ is estimated to be 0. In contrast, $\partial Q\,/\,\partial\delta$ is close to 0 and $\partial Q\,/\,\partial V_s$ would be $(2-V_t)\,/\,X$.

It is clear from Equations (8.14) and (8.15) that active power (P) is more sensitive to delta, whereas reactive power (Q) is affected by voltage magnitude. This is especially true when DG is integrated in a grid with a narrow phase angle δ. Based on the Equations (8.14) and (8.15), a system of control was developed for dispersing the true power used by the consumers from the wind turbines by managing the phase relationship of a power conversion technique and the reactive power generated by controlling the voltage amplitude v_t.

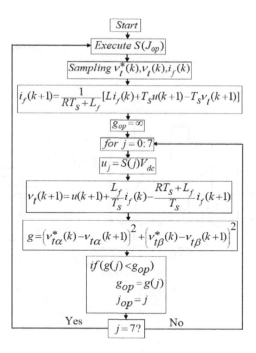

FIGURE 8.8   MPC algorithm flowchart

The power averaging approach is used to continuously measure the actual power consumption by the load. The amount of load power at any moment in time may be determined using the current drawn by the load and grid terminal voltage parameters shown below.

$$P_L = v_{ta} \cdot i_{la} + v_{tb} \cdot i_{lb} + v_{tc} \cdot i_{lc} \tag{8.20}$$

A low-pass filter can be used to obtain the actual power average ($P_{L,avg}$) using PL. The power generated by wind turbines is also monitored by Equation (8.21).

$$P_\omega = \frac{1}{2}\rho A v_\omega^3 = \frac{1}{2}\rho \pi v_\omega^3 R^2 C_p(\lambda, \beta) \tag{8.21}$$

Equation (8.22) specifies that the reference power P_{ref} which is the difference of wind power ($P\omega$) and average load power ($P_{L,avg}$), be supplied for source

$$P_{ref} = P_\omega - P_{L,avg} \tag{8.22}$$

The phase relationship δ of the voltage measured by the PCC is used to regulate the source and the wind's ability to share power effectively, while assuming a zero phase angle for the source voltage V_s. This takes into account changes in both the load and wind conditions. The net reference for the desired grid terminal voltage includes both the magnitude (v_{tref}) and the phase angle (δ).

When reference power $(P_{ref}/3)$ from Equation (8.22) is replaced in Equation (8.14) to get the phase angle online, the required instantaneous reference voltages may be represented as

$$v_{taref} = \sqrt{2}\, V_m \, \text{Sin}\left(\omega t + \delta\right)$$

$$v_{tbref} = \sqrt{2}\, V_m \, \text{Sin}\left(\omega t + \delta - 120^0\right) \qquad (8.23)$$

$$v_{tbref} = \sqrt{2}\, V_m \, \text{Sin}\left(\omega t + \delta + 120^0\right)$$

8.6 SIMULATION RESULTS

8.6.1 PERFORMANCE WITH MPC

Considering the number of loads connected across the feeder, the distribution system voltage should always be a balanced sinusoid between the line and ground. A non-linear diode bridge rectifier is coupled to the load with resistor capacitor integrated in parallel to emulate supply voltage disturbance. The assessed present wind power is not sufficient to compensate for the required power demand, i.e., $P_\omega < P_{L,avg}$. As a result of the diode bridge rectifier and feeder impedance, load voltages are distorted and fall much below their nominal value (400 V peak), as depicted in Figure 8.9. A shunt compensator is integrated with the network to mitigate the load voltage disturbances. At $t = 1$ s, we suppose that the wind speed has risen, i.e., $P > P_{L,avg}$. As can be seen from Figure 8.11, this causes a variation in reference power and, as a result, a drop in phase angle with regard to the source voltage. Consequently, additional wind power is supplied, and grid voltages are controlled to reference voltages of 400 V (peak), as is illustrated in Figure 8.10, with magnified corrected voltages in Figure 8.12.

8.6.2 ANALYSIS OF VOLTAGE SAG/SWELL

Sag/swell in source voltages should be examined as a specific scenario of power quality concerns, as shown in Figure 8.13. In a stable condition, the voltage peaks at 400 V but then experiences a dip for a few cycles at 0.5 s, and then in all phases of the supply voltage, there are voltage increase and dip events occurring at 0.8 s and 1.3 s, correspondingly. Consider a case in which the wind turbine gathers more energy than that of the mean power demand, i.e., $P > P_{L,avg}$, and the network is equipped along with a shunt controller that can supply harmonic current in addition to a portion of the consumer load demand. When the supply voltage experiences sag or swell, the phase angle delta undergoes a readjustment to establish a new value, which helps to mitigate the impact of voltage disruptions on the power supply. Figure 8.14 shows that the load voltages are both sinusoidal and balanced, and the shunt controller corrects any sag or swell, thereby preventing any impact on the load voltages.

8.6.3 ANALYSIS WITH UNEQUAL SOURCE VOLTAGES

During instabilities from the source side, it is critical that the management system have the ability to modulate the output voltages in order to maintain the proper balance.

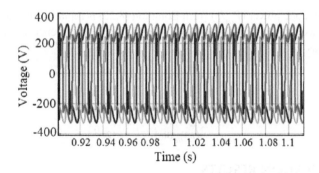

FIGURE 8.9   Grid terminal voltage without compensation

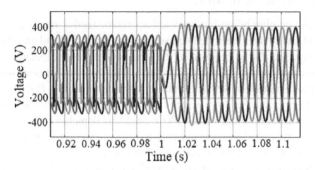

FIGURE 8.10   Grid terminal voltage with compensation

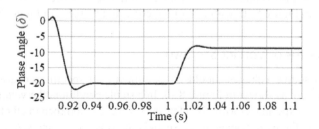

FIGURE 8.11   Phase angle

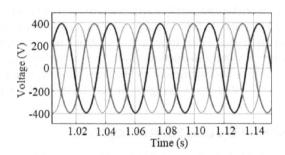

FIGURE 8.12   Enlarged image of the grid voltage

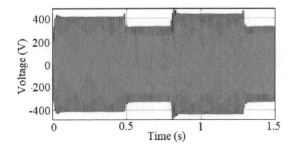

FIGURE 8.13   Source voltages

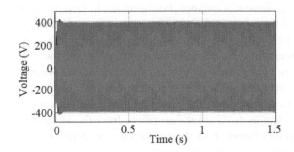

FIGURE 8.14   Grid terminal voltages

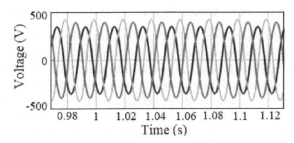

FIGURE 8.15   Unbalanced source voltages

To demonstrate this, consider the RMS voltages of 380 V, 400 V, and 420 V (peak voltage) for source voltages. This is presented in Figure 8.15. Consider the situation for which wind turbines capture more energy than the mean power consumption, i.e., $P > P_{L,avg}$, and the compensator is installed at $t = 1.0$ s. As observed in Equation (8.23), any imbalance in the supply voltages results in a fluctuation in the load power, leading to an adjustment in the phase angle delta to rectify the supply voltage imbalance. The load voltages become balanced sinusoids alongside overall harmonic distortion of approximately 2%, as shown in Figure 8.16.

8.6.4   ANALYSIS WITH MORE WIND POTENTIAL THAN LOAD DEMAND

Regardless of the disturbance from the load or supply side, a VSC-based shunt controller has the capacity to regulate the level of voltage across the grid-connected loads,

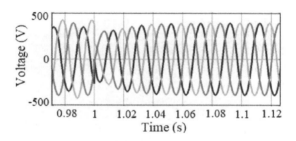

FIGURE 8.16 Compensated source voltages

thereby achieving sinusoidal balance. Furthermore, in response to fluctuations from both the grid and the wind source, the controller effectively manages power sharing between them. To test this, the wind velocity increased from 8 to 10 m/s at 1.5 s. As illustrated in Figure 8.17, we estimate that 23 kW of demand was split between both sources of energy prior to the rise in wind speed. It can be observed in Figure 8.18 that as wind speed increases, the phase angle steadily decreases in relation to the source; this causes a change in the reference power. In response to this, the shunt controller is efficient at distributing real power from the wind energy source while delivering remaining power from the source side. As demonstrated in Figures 8.19 and 8.20, the voltage at the grid end maintains the sinusoid in equilibrium.

8.6.5 Analysis with Less Wind Potential than Load Demand

During a reduction in wind speed, the exact opposite happens in this scenario. The speed of the wind is decreased from 10 m/s to 8 m/s at 1.5 s to test the controller's efficacy. As illustrated in Figure 8.21, we assume that 23 kW of demand was split between both energy sources prior to the reduction in wind speed. It can be observed from Figure 8.22 that as a wind speed decreases, the phase angle gradually rises in relation to the source, causing a shift in reference power. As a consequence, the source contributes more real power, whereas the remaining deficient power is provided by the wind. As is demonstrated in Figures 8.19 and 8.20, the shunt-based controller can manage grid terminal voltage to equilibrium sinusoids regardless of the low winds. In addition, as illustrated in Figure 8.21, the controller achieves proper power sharing between the source and the wind energy system.

8.6.6 Analysis During Load Variation

As depicted in Figure 8.23, the controller successfully achieved an equilibrium state with respect to a wind velocity of 6 m/s, leading to a higher contribution of real power from the source. The load is steadily raised starting at 1.5 s. As illustrated in Figure 8.24, as the load increases, the phase angle delta rises progressively in relation to the source and, consequently, the reference power changes. Therefore, as illustrated in Figure 8.23, the source delivers a higher amount of real power. As shown in Figures 8.19 and 8.20, the shunt controller's quick action ensures that increased load has no influence on the grid terminal voltages.

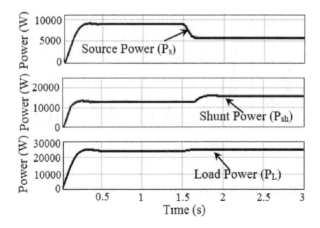

FIGURE 8.17 Power sharing across the grid terminal

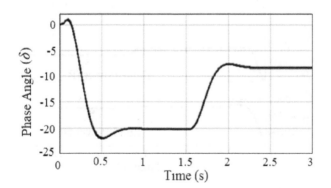

FIGURE 8.18 Phase angle steadily decreases due to increase in wind speed

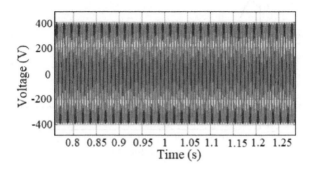

FIGURE 8.19 Load voltage during increased load

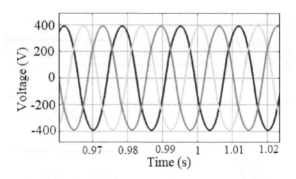

FIGURE 8.20   Magnified scale load voltage

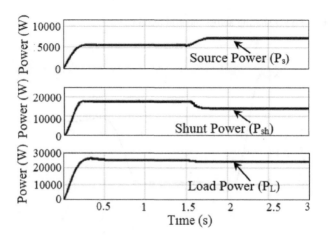

FIGURE 8.21   Power sharing across the grid terminal when wind speed decreases

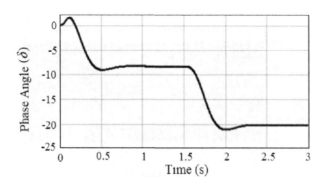

FIGURE 8.22   Phase angle gradually rises due to decrease in wind speed

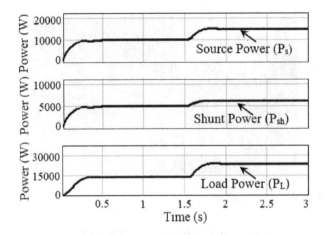

FIGURE 8.23 Power sharing across the grid terminal when load increases

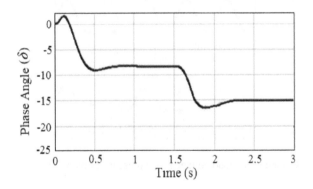

FIGURE 8.24 Phase angle

8.7 EXPERIMENTAL RESULTS

Figure 8.25 presents the experimental design of the single-phase power delivery system's architecture utilizing the dSPACE DS1103 environment to evaluate the effectiveness of the recommended control approach. The parameters used in both the MATLAB® software study and experimentation with the laboratory-based turbine emulator are outlined in Tables 8.1–8.4. The DC–DC converter is controlled to resemble the real wind turbine characteristics. The controller's efficacy is assessed using the following test circumstances.

8.7.1 ANALYSIS WITH NON-LINEAR LOAD AND IRREGULAR UTILITY VOLTAGE

The utility voltage is expected to be sinusoidal with an amplitude of 40 V (RMS), and it is used to power a load consisting of a non-linear diode bridge rectifier. Because of the existence of harmonic frequencies and the lack of shunt controller, the grid voltage becomes distorted, resulting in a decrease in its value by 20% (as depicted in Figure 8.26).

FIGURE 8.25   Experimental setup

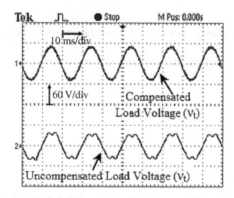

FIGURE 8.26   Compensated load voltage in the presence of harmonics

We consider that there is enough wind potential and that the controller is coupled to the network with the same load to analyze the system performance. Figure 8.26 shows that the shunt controller effectively compensates for the harmonic frequencies present in the non-linear load, resulting in balanced sinusoidal load voltage. Let's consider what happens when the supply voltage sags and swells, as indicated in Figures 8.27 and 8.28, respectively. It can be seen that swell and sag don't affect the load voltage since the shunt controller compensates for these supply voltage disturbances.

8.7.2   ANALYSIS WITH MORE WIND POTENTIAL THAN LOAD DEMAND

In order to study how the compensator reacts to different wind speeds, the wind speed is increased from 6 to 7 m/s, and the delta (phase angle) decreases in relation to the utility voltage. As a result, the shunt controller facilitates the delivery of additional real power from the wind source while distributing the remaining power between the wind and the grid. Figure 8.29 demonstrates that in order to produce a 200 W load, around 160 W is provided by the wind, and the balance of 40 W is provided by the source. It has been demonstrated that the controller has performed optimal wind power placement with the intention of reducing the load stress on the side of source.

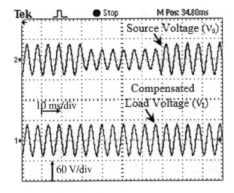

FIGURE 8.27   Load voltage compensation during sag in source voltage

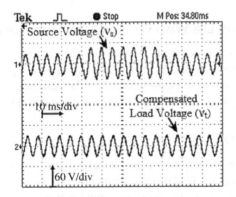

FIGURE 8.28   Load voltage compensation during swell in source voltage

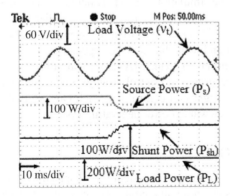

FIGURE 8.29   Power sharing across the grid terminal during more wind potential than load demand

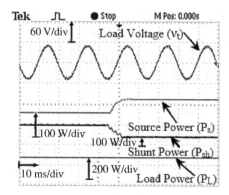

FIGURE 8.30 Power sharing across the grid terminal during less wind potential than load demand

8.7.3 ANALYSIS WITH LESS WIND POTENTIAL THAN LOAD DEMAND

As the speed of wind is reduced from 7 to 6 m/s in this scenario, the delta (phase angle) rises relative to the source voltage. As a result, the source power is greater than the power provided by wind. According to Figure 8.30, during the same load circumstances, 160W is provided from the source, and the remaining power is supplied by the wind.

8.8 CONCLUSION

Because of the rise in global power demand and pollution, wind energy technology is rising in prominence. As wind turbine output power is inconsistent, advanced control techniques are necessary to ensure that wind power export is compatible with the grid. To address this, an MPC method relying on a shunt controller based on a VSC has been explored to improve the distribution system's accuracy and energy efficiency. In view of the simulation and experimental studies, it is verified that in the existence of adequate wind potential, the shunt controller may adjust the grid voltage to an optimal sinusoid despite perturbations from the load or supply. Depending on the wind potential and load changes, the suggested control approach maximizes the power at the terminal of grid. MATLAB®/Simulink® simulation software is employed to study the proposed control strategy under unpredictable and varying wind conditions. The suggested control approach can be used to optimize energy effectively in hybrid renewable power generating systems.

8.9 FUTURE SCOPE

The importance of DG to the power system is anticipated to grow in the near future. Distributed generators can bring numerous benefits to the electrical market, including load reduction and demand response. Utilities can deploy DG to reduce peak load, provide complementary facilities such as voltage support, or reactive power, and improve power quality, particularly at the local level. This present study put forth a control technique for a variable-speed energy transformation system which is connected to

TABLE 8.1
Simulation model specifications for the proposed system

Description	Numerical Value
Feeder voltage	400 V (peak)
Frequency	50 Hz
Distribution line impedance	$(2 +j4)\ \Omega$
Filter capacitor	477.75 µf
Shunt branch impedance	$(1+j\ 2.54)\ \Omega$
DC coupler capacitor	4400 µf
DC coupler voltage V_{dc}	760 V

TABLE 8.2
Experimental model specifications for the proposed system

Description	Numerical Value
Utility voltage	40 V RMS, 50 Hz
Shunt branch resistance and inductance, R_f, L_f	10 Ω, 10 mH
Shunt filter capacitor, C_f	77.75 µF
Distribution line resistance and inductance, R_s, L_s	6.0 Ω, 25 mH
DC coupler voltage	170 V
Voltage measuring sensor	LEM LV25-P
Current measuring sensor	LEM LA25-NP

TABLE 8.3
Simulation model specifications for laboratory emulator

Description	Numerical Value
Generator rated capacity	20 kW
Normal speed	1500 rpm
Number of pole pairs	12
Normal voltage	400 V
Frequency	50 Hz
Turbine blade radius	4.0 m
Turbine inertia	50 kg m²
Coefficient of performance (C_p)	0.49
Wind velocity	7–12 m/s
Air density	1.2 kg/m³

the grid as a distributed resource. Wind power technology can improve distribution network reliability by injecting real power into the feeder to balance the load demand. Furthermore, it can give financial advantages by enhancing voltage quality and load balancing, thereby eliminating load congestion and potential blackouts.

TABLE 8.4

Experimental model specifications for laboratory emulator

Description	Numerical Value
Generator rated power	2.0 kVA
Rated current	2.8 A
Normal voltage	415 V
Poles pairs	4
Normal speed	1500 rpm
Generator constant	25 V s/rad/wb
Field winding flux	0.0156 wb
Winding inductance	0.0087 H
Winding resistance	1.565 Ω
Rated power of DC shunt motor	3 kW
Nominal voltage	230 V
Nominal current	13 A
Rated speed	1500 rpm
Armature winding resistance	3.5 Ω
Armature winding inductance	27 mH
Field winding resistance	220

REFERENCES

1. Rohten, J.A.; Espinoza, J.R.; Muñoz, J.A.; Pérez, M.A.; Melin, P.E.; Silva, J.J.; Espinosa, E.E.; Rivera, M.E. "Model predictive control for power converters in a distorted three-phase power supply". *IEEE Trans. Ind. Electron.*, Vol. 63, pp. 5838–5848, 2016.
2. Zhu, Ying; Hang, Jun; Zang, Haixiang; Zhao, Jingtao. "Sensor less HCS MPPT based control strategy for the DPF-WECS". In *2017 IEEE Energy Conversion Congress and Exposition (ECCE)*, 1–5 October, 2017.
3. Dursun, Emre Hasan; Kulaksiz, Ahmet Afsin. "MPPT control of PMSG based small-scale wind energy conversion system connected to DC-Bus". *Int. J. Emerg. Electr. Power Syst.*, Vol. 21(2), 1–13, 2020.
4. Terin, R.A.; Kaliannan, P.; Subramaniam, Umashankar; Thirumoorthy, Ad. "Power quality improvement of grid connected wind farms through voltage restoration using dynamic voltage restorer". *IJRER*, Vol. 6(1), 53–60, Jan. 2016.
5. El-Sattar, A.A.; Saad, N.H.; Shams El-Dein, M.Z. "Dynamic response of doubly fed induction generator variable speed wind turbine under fault". *Electr. Power Syst. Res.*, Vol. 78, pp. 1240–1246, Feb. 2008.
6. Mohan, M.; Panduranga Vittal, K. "Modelling and simulation of PMSG based wind power generation system". In *3rd IEEE Conference on Recent Trends in Electronics, Information &Communication Technology (RTEICT-2018)*, May 18–19, 2018.
7. Zhang, Z.; Wang, F.; Wang, J.; Rodriguez, J.; Kennel, R. "Nonlinear direct control for three-level NPC back-to-back converter PMSG wind turbine systems: experimental assessment with FPGA". *IEEE Trans. Ind. Inform.*, Vol. 13(3), pp. 1172–1183, 2017.
8. Chen, J.; Gong, C. "Constant-bandwidth maximum power point tracking strategy for variable speed wind turbines and its design details". *IEEE Trans. Ind. Electron.*, Vol. 60, pp. 5050–5058, 2013.

9. Mesemanolis, A.; Mademelis, C.; Kioskeridis, I. "Maximum electrical energy produc-tion of a variable speed wind energy conversion system". *IEEE Trans. Ind. Electron.*, Vol. 1, pp. 1029–1034, 2012.

10. Errami, Y.; Maaroufi, M.; Ouasssid, M. "Control scheme and maximum power point tracking of variable speed wind farm based on the PMSG for utility network connec-tions". In *International Conferenc on Complex Systems*, 2012.

11. Mendis, N.; Muttaqi, K.M.; Sayeef, S.; Perera, S. "Standalone operation of wind tur-bine-based variable speed generators with maximum power extraction capability". *IEEE Trans. Energy Convers.*, Vol. 27, pp. 822–834, 2012.

12. Yaakoubi, A.E.; Amhaimar, L.; Attari, K.; Harrak, M.H.; Halaoui, M.E.; Asselman, A. "Non-linear and intelligent maximum power point tracking strategies for small size wind turbines: Performance analysis and comparison," *Energy Rep.*, Vol.5, pp. 545–554, 2019.

13. Ararwal, V.; Agarwal, R.K.; Patidar, P.; Patki, C. "A novel scheme for rapid tracking of maximum power point in wind energy generation systems". *IEEE Trans. Energy Con-vers.*, Vol. 25, pp. 228–236, 2010.

9 Model Predictive Control of Multiple Renewable Energy Sources in Hybrid DC Microgrids for Power Flow Control

Apoorva Srivastava and R.S. Bajpai

9.1 INTRODUCTION TO DISTRIBUTED ENERGY RESOURCES

Due to the increase in DC loads, DC microgrids are becoming more prominent. Considering DC system efficacy, economy, and scale, DC microgrids offer considerable benefits over AC microgrids. The concept of a DC microgrid has a significant advantage in that it allows for the incorporation of many forms of sustainable sources that are primarily DC sources, such as hydrogen fuel cells (FCs) and solar energy, as well as compact gas turbines, computers, fluorescent lights, variable speed drives, and a wide range of other business and household equipment and hardware. As indicated in Table 9.1, these home appliances alone consume a large amount of energy. The stable functioning of a DC microgrid necessitates the incorporation of an efficient control strategy for voltage and current sharing between several renewable energy sources. To ensure the consistent functioning of a DC microgrid, alternate energy sources must be controlled close to their most efficient point, irrespective of the external conditions [1].

During the last decade, numerous load distribution approaches have been studied in the literature. Sources connected in parallel [2], cascaded load distribution techniques [3], local control techniques [4], islanded modes of function [5], power sharing techniques based on voltage droop [6] or controlling load frequency [7], and self-adaptive inertia regulation [8] are some studied methods that have been identified for sharing power [9]. The control strategy, on the other hand, necessitates a communication network among modules, which raises system costs and minimizes system reliability. In order to overcome above challenges, a distributed control strategy has been suggested to assure corresponding load sharing in a low-voltage DC microgrid.

9.2 DC MICROGRID

Figure 9.1 demonstrates the DC microgrid suggested in this study. Table 9.1 shows the parameters of sustainable energy sources. Hybrid power generation schemes focusing on sustainable energy sources have recently shown to be promising due to

DOI: 10.1201/9781003311829-9

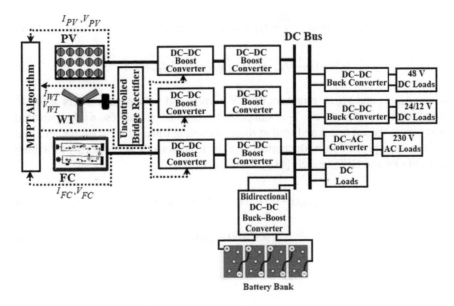

FIGURE 9.1 DC microgrid with distributed generation resources

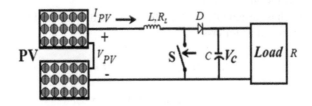

FIGURE 9.2 DC–DC boost converter with photovoltaics

their energy efficiency and consistent power delivery in remote locations far from grids. Microgrids are designed to operate independently, with the incorporation of renewable energy sources benefiting utility providers and providing customers with increased efficacy and power quality. Hybrid microgrids consist of battery storage systems (BSS), photovoltaic sources (PV), wind turbines (WT), proton exchange membrane fuel cells (PEMFC), electric vehicles (EVs), and AC and DC loads. The DC microgrid receives the generated electricity from renewable energy sources and distributes it to diverse loads spread all across the DC microgrid.

9.3 DC MICROGRID MODEL

DC–DC boost converters are used to transfer power from various energy sources into the DC microgrid. Figure 9.2 depicts an analogous arrangement of a solar power device with a boost converter. For FC and WT, a similar converter configuration is employed. The mathematical representation of a discrete-time DC–DC power converting device can be expressed as

$$I_{PV}(k+1)=(1-\frac{T_s R_L}{L})I_{PV}(k)+(u(k)-1)\frac{T_s}{L}V_C(k)+\frac{T_s}{L}V_{PV}(k) \qquad (9.1)$$

$$V_C(k+1)=\frac{T_s}{C}I_{PV}(k)+(1-\frac{T_s}{CR})V_C(k)-\frac{T_s}{C}I_{PV}(k)u(k) \qquad (9.2)$$

Where $I_{PV}(k)$ and $V_{PV}(k)$ represent the PV current and voltage parameters at the kth sampling time, respectively. T_s is sampling time, and $V_C(k)$ represents the potential exist across the capacitor and load at the kth sampling time. The PV source, FC, and WT can be expressed in mathematical form by Equations (9.3–9.6) during switching states of ON and OFF, i.e., (S={1,0}), correspondingly.

$$I_{rzt,0}(k+1)=I_{rzt}(k)+\frac{T_s}{L}(V_{rzt}(k)-V_C(k)) \qquad (9.3)$$

$$V_{rzt,1}(k+1)=(1-\frac{T_s}{RC})V_C(k) \qquad (9.4)$$

$$I_{rzt,1}(k+1)=\frac{T_s}{L}V_{rzt}(k)+I_{rzt}(k) \qquad (9.5)$$

$$V_{rzt,0}(k+1)=\frac{T_s}{C}I_{rzt}(k)+(1-\frac{T_s}{CR})V_C(k) \qquad (9.6)$$

Where $s=$ FC, $r=$ PV, and $t=$ WT.

When the switch is turned ON or OFF, the forecasted values of FC, solar, and WT power can be estimated as follows:

$$P_{rzt,1}(k+1)=I_{rzt,1}(k+1)\cdot V_{rzt,1}(k+1) \qquad (9.7)$$

$$P_{rzt,0}(k+1)=I_{rzt,0}(k+1)\cdot V_{rzt,0}(k+1) \qquad (9.8)$$

The controller's generalized cost function is then minimized in accordance with the following [10]:

$$J_{rzt,1}=\left|P_{rzt,1}(k+1)-P_{rzt,ref}\right| \qquad (9.9)$$

$$J_{rzt,0}=\left|P_{rzt,0}(k+1)-P_{rzt,ref}\right| \qquad (9.10)$$

As illustrated in Figure 9.3, the switching signal with the optimal cost is chosen and supplied to the converter using Equations (9.9) and (9.10).

9.4 FUEL CELL MODEL

Fuel cells have demonstrated effectiveness in generating power with minimal emissions and high efficiency, and their adaptable design makes them an attractive alternative to be used as a dispersed energy resource. In [11], researchers took into consideration the dynamic stack model (PEMFC). The electrolytic reaction of hydrogen and oxygen in a PEMFC transforms chemical energy into electricity. Water formed as a residue of energy conversion is fed into an electrolyzer to make

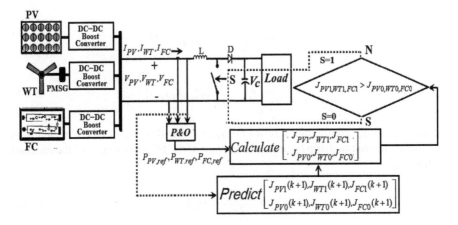

FIGURE 9.3 Maximum power extraction scheme

hydrogen. One of the key variables is U_f, which may be accessed in an FC and reacts electrochemically as shown in Equation 9.11.

$$U_f = \frac{q_{H_2}^{in} - q_{H_2}^{out}}{q_{H_2}^{in}} = \frac{q_{H_2}^r}{q_{H_2}^{in}} \tag{9.11}$$

The molar flow of hydrogen is represented by qH_2. High hydrogen utilization is essential for small-scale devices employing pure hydrogen as a fuel. Electrical output of the FC is affected by regulating the rate of flow into the FC. Increasing the usage factor, on the other hand, results in a significant voltage decrease. The best utilization of fuel, $U_{f,opt}$, is kept at approximately 85 percent, so that $q_{H_2}^r$ is directly related to the FC stack current I_{FC}, allowing the device to operate at maximum efficiency. The FC stack's typical voltage parameter is calculated by utilizing Ohm's law and Nernst's equation. The FC stack potential is modeled using the expressions provided herein.

$$V_{FC} = N_0 \left[E_0 + \frac{RT}{2F} \left(\ln \left\{ \frac{P_{H_2} P_{O_2}^{0.5}}{P_{H_2O}} \right\} \right) \right] - r I_{FC} \tag{9.12}$$

where E_0 denotes the voltage related with the chemical free energy; N_0 specifies how many cells are in a series.; T refers the temperature; R indicates the universal gas constant; I_{FC} refers the FC stack current; and F represents the Faraday's constant. The differential equations that determine P_{H_2}, P_{H_2O}, and P_{O_2} are as follows.

$$P_{H_2} = -\frac{1}{t_{H_2}} \left[P_{H_2} + \frac{1}{K_{H_2}} \left\{ q_{H_2}^{in} - 2K_r I_{FC} \right\} \right] \tag{9.13}$$

$$P_{H_2O} = -\frac{1}{t_{H_2O}}\left[P_{H_2O} + \frac{2}{K_{H_2O}}K_r I_{FC}\right] \qquad (9.14)$$

$$P_{O_2} = -\frac{1}{t_{O_2}}\left[P_O + \frac{1}{K_{O_2}}\left\{q_{O_2}^{in} - K_r I_{FC}\right\}\right] \qquad (9.15)$$

Where $q_{O_2}^{in}$ and $q_{H_2}^{in}$ are the molar flow of oxygen and hydrogen, and the constant K_r is dictated by the relationship between the current of the FC and the rate of the reactant hydrogen.

$$q_{H_2}^r = \frac{N_o I_{FC}}{2F} = 2K_r I_{FC} \qquad (9.16)$$

The FC's functional limit is managed in the following way.

$$\frac{0.7q_{H_2}^{in}}{2K_r} = I_{FC,min} \le I_{FC,ref} \le I_{FC,max} = \frac{0.9q_{H_2}^{in}}{2K_r} \qquad (9.17)$$

9.5   MAXIMUM POWER POINT TRACKING CONTROL

This research aims to combine the benefits of the commonly used incremental conductance (INC) framework with model predictive control (MPC). The utilization of proposed maximum power point (MPP) tracking through MPC satisfies two critical requirements, dependability and speed, while preventing unwanted oscillations near MPP. Furthermore, the control technique ensures that maximum power is extracted from sustainable power sources.

To transfer maximum power to the DC microgrid, DC–DC boost converters are incorporated, as illustrated in Figure 9.3. A DC–DC boost converter has two stages. Maximum power is captured in the first stage, and output voltage is controlled in the second stage, which is essential for power flow regulation. As illustrated in Figure 9.4, incremental conductance – maximum power point tracking (INC MPPT) examines relevant references for MPC that govern further switching signals for the suggested objective functions.

To command the DC–DC converters, optimum switching states are evaluated based on projected voltage and current values. Figure 9.5 displays the recommended MPC power-harnessing technique for minimizing the cost function. Figures 9.6 and 9.7 show the experimental findings gathered using the proposed MPC approach as well as the standard incremental conductance method, respectively.

The system is tested at varying irradiance levels, with the irradiance level starting at 700 W/m² and subsequently increasing to 800 W/m². The effectiveness of MPC under variable irradiance is superior to that of the conventional INC technique, as shown in Figures 9.6 and 9.7.

Figures 9.8 and 9.9 show the experimental findings derived according to the proposed MPC employing WT and FC technology. To investigate the performance, the flow of hydrogen fuel and the velocity of wind are raised to 6 L/min from 4 L/min

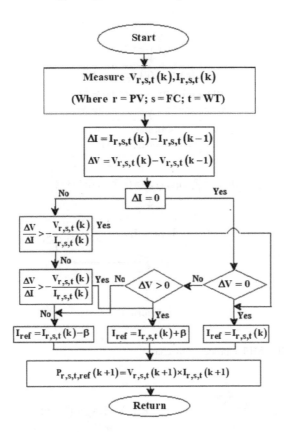

FIGURE 9.4 INC algorithm flowchart

and to 7 m/s from 6 m/s, respectively. It can be seen that in comparison to the conventional INC MPPT, the proposed MPC technique yields reduced equilibrium state disturbances and improved tracking reaction time.

9.6 UTILIZING MPC FOR POWER FLOW REGULATION

Droop control and its derivatives are extensively used in DC microgrid power sharing control. Although droop control can eliminate the need for communication lines, it is not without its drawbacks. These include voltage fluctuations and the inability to accurately share power because of output impedance uncertainty. In a DC microgrid, achieving proper load sharing among the decentralized renewable energy sources is a major goal. Cable resistance produces significant voltage droop in low-voltage DC systems, affecting the sharing of power. An MPC controller incorporating a discrete-time Kalman filter can be created to achieve proportionate power sharing by modifying controlling variables to the necessary form and minimizing steady-state deviations between storage batteries and renewable sources.

Figure 9.10 shows a block layout of the suggested power sharing control scheme. The purpose of MPC is to minimize objective function J in addition to accounting for the issue constraints that lie within an anticipated control frame N.

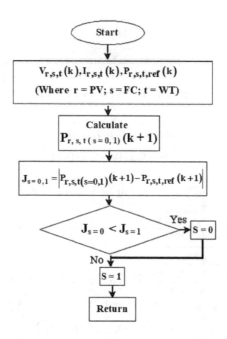

FIGURE 9.5   MPC-MPPT process flowchart

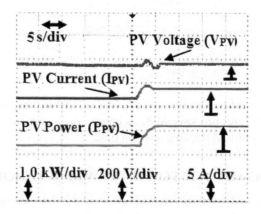

FIGURE 9.6   "PVystem" parameters with MPC-MPPT

$$P_m = P_G - P_L \tag{9.18}$$

$$P_m = P_{WT} + P_{PV} + P_{FC} - P_L \tag{9.19}$$

where P_L stands for demand power, P_G stands for produced power, P_m stands for total accessible power, P_{PV} stands for photovoltaic power, P_{WT} stands for wind-turbine-generated power, and P_{FC} stands for FC power. A switching system is designed with the primary aim of controlling the load current to specific set points. Furthermore, the

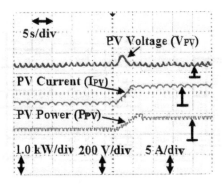

FIGURE 9.7   PV system parameters with INC

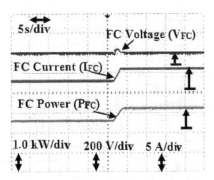

FIGURE 9.8   Fuel cell system parameters with MPC-MPPT

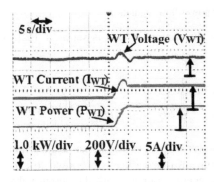

FIGURE 9.9   Wind turbine system parameters with MPC-MPPT

system should be robust and steady in the presence of unexpected variations from the source or load sides.

The control scheme is defined as follows, ignoring switching losses.

$$I_{L,error}(k) = I_{L,ref} - I_L(k) \tag{9.20}$$

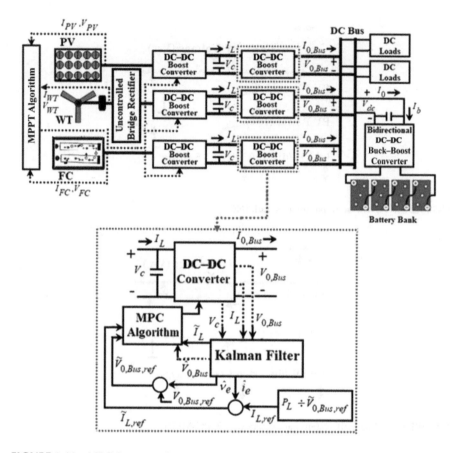

FIGURE 9.10 MPC for power flow regulation

A reference current for the controller loop is set as in [12] to achieve the desired results.

$$I_{L,ref} = \frac{V_c}{2R_L} - \sqrt{\left(\frac{V_c}{2R_L}\right)^2 - \frac{V^2_{0,Bus,ref}}{RR_L} + \left(V_{0,Bus,ref} - V_{0,Bus}\right)} \qquad (9.21)$$

The load demand in power distribution systems is uncertain. An MPC controller assisted by a discrete-time Kalman filter is proposed to achieve respective power sharing from renewable energy sources, with the output voltage and current set for maximum efficiency, accounting for variations in power from either the source or load side. To attain the maximum power output, this chapter intends to combine the benefits of the INC methodology alongside MPC for maximum power-harnessing control. The use of incremental conductance along with MPC satisfies two critical requirements, speed and dependability, while preventing unwanted oscillations near MPP. Furthermore, the control technique ensures that maximum power is extracted from renewable energy sources.

As illustrated in Figure 9.10, current and voltage estimates $\hat{v}_e$ and $\hat{i}_e$ are unveiled, allowing the error in the output voltage to be eliminated in the event of load fluctuations. The performance of MPC under different load conditions is explored, and it is discovered that the Kalman filter takes care of the predicted current and voltage states throughout differing loads for the purpose of giving the necessary load power. Energy storage batteries are operated to deliver appropriate power when energy generated from clean energy resources, but this is insufficient to complement the load demand if failure in the equipment occurs, due to system constraints.

Control variables can be formulated as follows.

$$\tilde{V}_{0,Bus,ref} = V_{0,Bus,ref} - \hat{V}_e \tag{9.22}$$

$$\tilde{I}_{L,ref} = I_{L,ref} - \hat{i}_e \tag{9.23}$$

The determined state parameters in conjunction with external influences are

$$x_\eta = \begin{bmatrix} I_L & V_{0,Bus} & i_e & v_e \end{bmatrix}^T \tag{9.24}$$

By solving the equations for load flow for the mth source converters and n^{th} time-varying loads, the cost function is minimized to increase system reliability due to the loads [13].

$$\sum_m I_{Lm} = \sum_n \frac{P_{Ln}}{\tilde{V}_{0,Bus,ref}} \tag{9.25}$$

where P_{Ln} is the nth variable load power and I_{Lm} is the mth source converter current.

The cost function formulation is critical in MPC performance because it limits the fluctuation from the desired value ($V_{0,Bus,ref}$ and $I_{L,ref}$) and may be defined as:

$$J_{s=\sigma}^{\sigma=0,1} = \lambda_1 \left| V_{c,s=\sigma}(k+1) - V_{0,Bus,ref} \right| + \lambda_2 \left| I_{L,s=\sigma}(k+1) - I_{L,ref} \right| \tag{9.26}$$

where λ_1 and λ_2 are $[1/V]$ and $[1/A]$ respectively.

9.7 KALMAN FILTER DESIGN

The Kalman filter is a method for determining unknown variables by utilizing observations collected over time. This technique has found broad use in numerous applications. The system's state space model can be expressed as

$$x(k+1) = A x(k) + B u(k) + \omega(k)$$
$$y(k) = C x(k) + v(k) \tag{9.27}$$

where $u_k \in R^n$, $x_k \in R^n$, $y_k \in R^p$, ω_k, and v_k are zero-mean white noise processes and have known covariance matrices Q_k and R_k.

$$\omega_k = N(0, Q_k)$$
$$v_k = N(0, R_k)$$

(9.28)

They are also described as

$$E\begin{bmatrix} \omega_k \omega_k^T & \omega_k v_k^T \\ v_k \omega_k^T & v_k v_k^T \end{bmatrix} = \begin{bmatrix} Q_k & 0 \\ 0 & R_k \end{bmatrix}$$

(9.29)

The formulation for the Kalman filter structure is presented in [14].

$$\tilde{x}_{k+1|k} = A_k \tilde{x}_{k|k-1} + B_k u_k + K_F (y_k - C_k \tilde{x}_{k|k-1})$$

(9.30)

$$P_{k+1|k} = A_k P_{k|k} A_k^T + K_k Q_k K_k^T$$

(9.31)

$$P_{k|k} = \begin{bmatrix} I - K_F C_k \end{bmatrix} P_{k|k-1}$$

(9.32)

$$K_F = P_{k|k-1} C_k^T (C_k P_{k|k-1} C_k^T + R_k)^{-1}$$

(9.33)

Where K_F is Kalman gain, $y(k) = v_0(k)$, and $x(k) = I\begin{bmatrix} I_L(k) & V_c(k) \end{bmatrix}^T$.

9.8   SIMULATION RESULTS

DC–DC converters are essential for maintaining a consistent voltage source to loads associated with a DC microgrid. It is expected that when generated power is inadequate to meet power demands, battery storage is triggered to make up the difference, depending on system constraints. To assess the effectiveness of the MPC approach in different circumstances, a simulation experiment is conducted. Parameters of the DC microgrid under study are shown in Table 9.1.

9.8.1   Power Distribution for Increased Generated Power and Load

The results of the DC microgrid experiments are shown in Figure 9.11 (a–d). Figure 9.11 (a) and (b) show the power produced by FC, PV, and WT sources with and without the implementation of Kalman filters, respectively. As illustrated in Figure 9.11 (a–b), power output fluctuates depending on power gathered from multiple resources at time $t = 4$ s, $t = 5$ s, and $t = 7$ s. Figure 9.11 (c) and (d) show that the load power increases from 15 kW to 20 kW at $t = 4$ s and 5 s. Because the maximum power provided by FC, solar, and WT systems is insufficient to meet the load requirements, stored energy contributes to the deficient power, as illustrated in Figure 9.11

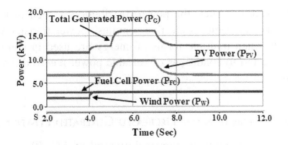

FIGURE 9.11 (a) Power sharing strategy with a Kalman filter

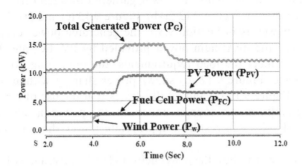

FIGURE 9.11 (b) Power sharing strategy with and without a Kalman filter

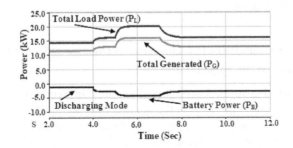

FIGURE 9.11 (c) Power balancing with a Kalman filter

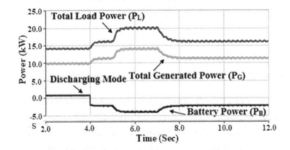

FIGURE 9.11 (d) Power balancing with and without a Kalman filter

(c–d). Without a Kalman filter, the power distribution is seen in Figure 9.11 (d). Figures 9.11 (a) and (c) demonstrate that seamless power transfer with improved quality is possible when multiple forms of energy are used in conjunction with a Kalman filter subjected to conditions of variable power received from the demand and supply ends.

9.8.2 POWER DISTRIBUTION FOR DECREASED GENERATED POWER AND LOAD

Figure 9.12 (a) and (b) show the power delivered by FC, PV, and WT sources with and without Kalman filters. During time $t = 4$ s, $t = 5$ s, and $t = 7$ s, power generated changes in line with the power gathered through different forms of energy, as illustrated in Figure 9.12 (a–b). Figure 9.12 (c) and (d) show that with the same load power as in Section 9.8.1, the generated power drops at $t = 4$ s and 5 s. Because the maximum power provided by PV, FC, and WT sources is insufficient to meet the load requirements, the remaining power is shared by energy storage. Figure 9.12 (d) shows power sharing without the use of a Kalman filter. Figure 9.12 (a) and (c) illustrates that the use of a Kalman filter in controlling multi-energy sources under varying power instances from both supply and load sides can result in appropriate power sharing and enhanced power quality.

Figure 9.12 (a) and (b) show the power delivered by FC, PV, and WT sources with Kalman filters and without. During time $t = 4$ s, $t = 5$ s, and $t = 7$ s, power generated changes in line with the power gathered through different forms of energy, as illustrated in Figure 9.12 (a–b).

9.8.3 POWER DISTRIBUTION FOR DIFFERENT SCENARIOS OF GENERATED POWER WITH THE SAME LOAD

Under the identical load circumstances illustrated in Sections 9.8.1 and 9.8.2, Figure 9.13 (a) shows an increase in the amount of power developed at time $t = 5$ s in accordance with the power gained through multiple sources of clean energy. Since the optimum power produced from multiple clean energy sources cannot satisfy the power requirements of the load, the remaining power is shared by the energy storage system, as depicted in Figure 9.13 (b). The consistency of coordination across power generating and delivery systems is observed in the numerous instances outlined above, demonstrating the tracking control system's robustness.

9.8.4 USING POWER POOLING FOR 24-HOUR OPERATION

The maximum power generation conditions during "9:00 to 15:00" are depicted in Figure 9.14 (a) using the solar insolation parameters and estimated speed of wind. The energy from sustainable energy resources used to power the load demand, as well as the battery power charge and discharge, is depicted in Figure 9.14 (b) for the duration of 24 hours. Figure 9.14 (b) shows that power generated is adequate to fulfill

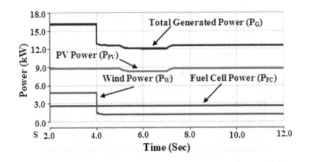

FIGURE 9.12 (a) Power balancing with a Kalman filter

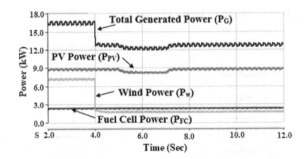

FIGURE 9.12 (b) Power balancing with and without a Kalman filter

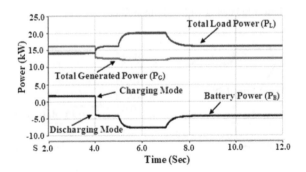

FIGURE 9.12 (c) Power balancing with a Kalman filter

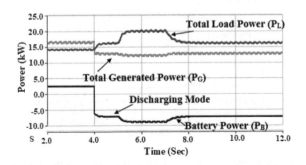

FIGURE 9.12 (d) Power balancing with and without a Kalman filter

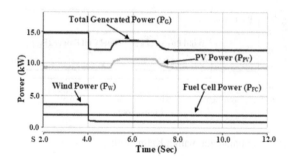

FIGURE 9.13   (a) Power balancing with a Kalman filter

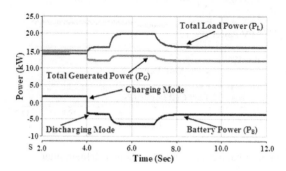

FIGURE 9.13   (b) Power balancing with a Kalman filter

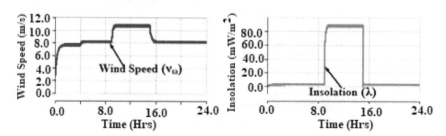

FIGURE 9.14   (a) Wind and solar parameters for 24 hours of operation

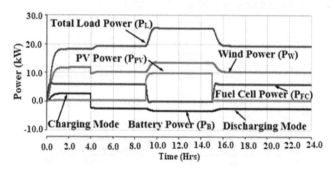

FIGURE 9.14   (b) Power sharing from multiple energy sources with a Kalman filter

load demand during the first four hours, and the control system enables charging operations based on the charging level of the storage batteries.

The Kalman filter changes the power generation pointers for solar, FCs, and wind, such that the control system can handle the load's insufficient power and charge or discharge the storage battery to satisfy the energy requirement.

9.9 EXPERIMENTAL RESULTS

A scaled-down experiment in the lab is established to validate the efficacy of the suggested controlling approach on the basis of MPC. The experimental setup uses a permanent magnet synchronous generator connected to a DC shunt motor to imitate the characteristics of a WT. Like in windmills, characteristics are generated utilizing a microcontroller and an armature driven DC shunt motor. Tables 9.2 and 9.3 describe the parameters of renewable resources and the DC–DC converter for the laboratory investigation. Erasable programmable ROM (EPROM), DSpace DS1103, TMS 320F240 DSP, a TMDSEVM572X assessment panel, static RAM, and a Mitsubishi PM50CL1A060 energy module with sensors are used to run the control algorithm. Figure 9.18 depicts the hardware configuration that is being tested for the proposed research, along with the instruments for experimentation for the suggested hybrid DC microgrid. Switching signals for the boost converters are generated using pulse width modulation (PWM) method at a 5 kHz frequency. The control module is built using MATLAB Simulink control block sets assembled in an m-file and linked to a real-time framework.

The voltage and current measurements are calculated using Hall effect sensors (LEM LV25-P) and (LEM LA25-NP). Using an anticipated horizon sequence ($N = 1$), the suggested MPC approach is executed every $T_s = 10$ s. The controller performance is demonstrated by the following experimental findings.

TABLE 9.1
Variables for decentralized generation

Description	Value
PV system rating	25 kW
Optimum voltage at MPP (V_{mpp})	17.8 V
Maximum current at (I_{mpp})	5.58 A
Voltage without load (V_{oc})	22.0 V
Maximum current during short circuit (I_{sc})	8.03 A
Fuel cell system capacity (PEMFC)	15 kW
Capacity of wind turbine system	25 kW
Maximum value of wind speed (v_w)	12.0 m/s
Air density (ρ)	1.2 kg/m^3
Coefficient of performance (C_p)	0.46

TABLE 9.2

Hardware design system rating

Description	Value
PV system capacity (250 W, Ea-16)	4 kW
Optimum voltage (V_{mpp})	17.5 V
Maximum current (I_{mpp})	8.87 A
Voltage without load (V_{oc})	24.3 V
Maximum current allowed during short circuit (I_{sc})	9.40 A
Fuel cell system (PEMFC)	4 kW
Fuel cell voltage at V_{mpp}	52.0V
Fuel cell current at I_{mpp}	45.6A
Wind turbine system capacity	4 kVA, 410V
No. of poles pairs	6
Rated speed	1500 rpm
Generator constant	27V s/rad/wb
Magnetizing flux	0.0132 Wb
Generator winding inductance	0.0056 H
Generator winding resistance	1.675 Ω
Rating of DC shunt motor	250V DC
Storage battery bank (12 V, 180 AH, Ea-04)	48V DC

TABLE 9.3

Hardware design components and their ratings

Hardware Components	Rating	Value
Boost converter (WT)	Input: 45–90 V DC, Output: 300–400 V DC	4 kW
Boost converter (PV)	Input: 45–90 V DC Output: 300–400 V DC	3.5 kW
Boost converter (FC)	Input: 40–85 V DC, Output: 300–400 V DC	3.5 kW
Boost converter (Ea-03)	Input: 390–420 V DC, Output: 300–400 V DC	3.5 kW
Step-down converter	Input: 390–420 V DC, Output: 52 V DC	3.5 kW
Step-down converter	Input: 390–420 V DC, Output: 12/24 V DC	3.5 kW
Bidirectional boost converter	Input: 390–420 V DC, Output: 52 V	3.5 kW
Bridge rectifier (WT)	700 V, 5 A	3.5kW

9.9.1 FUNCTIONING OF A MICROGRID UNDER AN INCREASED LOAD

Figure 9.15 (a) shows that the initial load is fixed at 4.1 kW, and the power provided through WT, PV, and FC systems is 1.2 kW, 1.8 kW, and 1.1 kW, respectively, to meet this requirement. Then, the load is raised to 5.0 kW, leading to a shift in the pooling of power such that WT, PV, and FC resources are 1.6, 2.0, and 1.4 kW, respectively. The Kalman filter modifies current and voltage reference to the point where

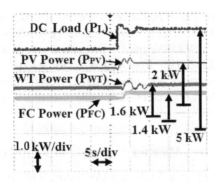

FIGURE 9.15 (a) Power sharing across several renewable energy sources using a Kalman filter

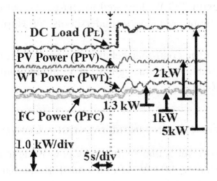

FIGURE 9.15 (b) Current sharing across several renewable energy sources using a Kalman filter

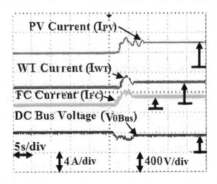

FIGURE 9.15 (c) Power sharing across several renewable energy sources without a Kalman filter

the controller meets the load's power requirements. During the transient, the DC output voltage falls and then returns to the usual range of 400 V. Current and voltage sharing waveforms from WT, PV, and FC generators are shown in Figure 9.15 (b). When the load is raised to 5.0 kW from 4.1 kW, Figure 9.15 (c) shows the behavior of the controller in the absence of a Kalman filter. The performance of the power supplied to loads through WT, PV, and FC renewable resources is deteriorated, as can be observed from Figure 9.15 (c).

9.9.2 Functioning of a Microgrid under a Reduced Load

Load Scenario is reduced from 5 kW to 3 kW. Figure 9.16 (a) depicts the outcomes of power sharing. The Kalman filter modifies current and voltage values to the point where the controller's steady-state error function is minimized. The overall shared load from WT, PV, and FC was found to be 1.0 kW, 1.2 kW, and 1.0 kW, respectively, with an extra 200 W delivered to charge the battery system. DC voltage is maintained without overshoots, undershoots, or ripples, as demonstrated Figure 9.16 (b).

Figure 9.16 (c) shows the controller behavior without the Kalman filter. The efficiency of power delivered to loads through WT, PV, and FC sources has deteriorated, as shown in Figure 9.16 (c).

9.9.3 Microgrid Functioning in the Event of a Converter Failure

In this case, the FC converter is unable to supply power due to an increase in power consumption from 2.0 kW to 4.0 kW. After a load shift, the Kalman filter attends to the predicted current and voltage values to provide the appropriate power demand. The power delivered by the WT and PV sources is raised to complement the inadequate power, as shown in Figure 9.17. As compared to the control technique used in [15], it is worth noting that the suggested controller has superior tracing capability.

9.10 CONCLUSION

The use of traditional assets, for example fossil fuels and coal, has increased significantly as a result of an upsurge in demand for supplies throughout the world. However, these conventional resources are known to cause environmental pollution, prompting researchers to explore alternative methods of meeting the world's energy demands using a variety of readily available energy resources.

The increasing global need for power has led to a major increase in the use of traditional resources, which have a track record of generating environmental damage. Because of this, scientists are looking into other approaches to produce electricity from inexhaustible energy sources. Among these sources, solar and wind energy are becoming increasingly popular due to their availability in nature, high capacity, and potential to meet the rising power demand. DC-powered microgrids are superior to microgrids with AC power in a number of ways, including system reliability, affordability, and design component reduction. Furthermore, as many commercial and residential

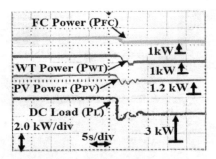

FIGURE 9.16 (a) Power sharing across several renewable energy sources with a Kalman filter

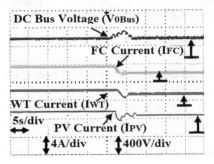

FIGURE 9.16 (b) Current sharing across several renewable energy sources with a Kalman filter

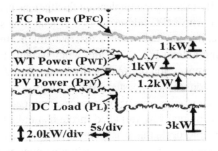

FIGURE 9.16 (c) Power sharing across several renewable energy sources without a Kalman filter

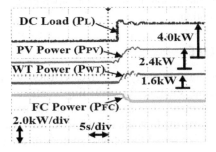

FIGURE 9.17 Power sharing across several renewable energy sources with a Kalman filter

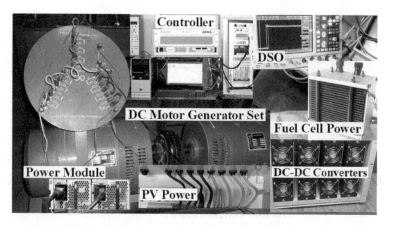

FIGURE 9.18 Experimental setup

products and equipment need DC power, the usage of microgrids based on DC is favorable. However, since renewable energy sources are irregular, incorporating them into DC power grids presents technological challenges. An appropriate control method has been presented to handle these issues and enhance the entire distribution system's functionality as a whole. A distribution network that integrates multiple distributed generation systems such as battery storage systems, WT, PV, and FC is called a hybrid microgrid. An MPPT control mechanism is necessary to optimize the functioning of green energy technologies. The objective of this chapter was to combine the advantages of the INC with MPC. By minimizing unpleasant fluctuations near MPP, the inclusion of INC MPPT together with MPC addresses two crucial issues, speed and dependability. An MPC device, in conjunction with a discrete-time Kalman filter, is designed to enable energy sharing among FC, solar, and wind systems to regulate the output voltage and current references for optimal performance based on power variations from the load or source sides. The suggested MPC strategy controls power distribution among the battery energy storage, renewable sources, and loads. The effectiveness of the suggested control technique was validated through simulations and experimental tests conducted on a DC microgrid that was incorporated with a battery storage system and renewable energy sources. The proposed technique may provide efficient energy distribution across DG units throughout a DC microgrid, enabling quick and precise monitoring control. A combination of AC and DC microgrids could serve as another possible energy supply option.

9.11   FUTURE SCOPE

The integration of distributed energy resources (DERs) with traditional systems emerges as an intelligent solution for providing continuous and secure power, even during peak load demand. The escalating need for power generation and the depletion of fossil fuels has made it imperative to harness energy from renewable resources. Microgrids are powered by reliable sources of clean energy that include FCs, wind, and solar photovoltaics. The introduction of DC microgrids eliminates the need for traditional AC grids. The hybrid of DC and AC microgrids will offer additional benefits to consumers at various levels. This chapter proposes the modeling and design of a hybrid DC microgrid that can reduce numerous power conversions by making efficient use of renewable energy sources.

REFERENCES

1. H. U. Rahman Habib, S. Wang, M. R. Elkadeem, and M. F. Elmorshedy, "Design Optimization and Model Predictive Control of a Standalone Hybrid Renewable Energy System: A Case Study on a Small Residential Load in Pakistan", *IEEE Access*, Vol. 7(1), 2019.

2. Qinjin Zhang, X. Zhuang, Y. Lu, C. Wang, and H. Guo, "A Novel Autonomous Current Sharing Control Strategy for Multiple Paralleled DC-DC Converters in Islanded DC Microgrid", *Energies, MDPI*, Vol. 12(20), 3951, 2019.

3. Leony Ortiz, Rogelio Orizondo, Alexander Águila, Jorge W. González, Gabriel J. López, and Idi Isaac, "Hybrid ac/dc Microgrid Test System Simulation: Grid Connected Mode Heliyon", *Science Direct*, Vol. 5(12), 1–21, 2019.

4. Jaynendra Kumar, and Nitin Singh, "Design, Operation and Control of a Vast DC Microgrid for Integration of Renewable Energy Sources", *Renewable Energy Focus*, Vol. 34(1), 17–36, 2020.

5. Apoorva Srivastava, and R. S. Bajpai, "Model Predictive Control of Grid Connected Wind Energy Conversion System", *IETE Journal of Research*, Vol. 66(1), 2020.

6. Zeinab Karami, Qobad Shafiee, Yousef Khayat, Meysam Yaribeygi, Tomislav Dragičević, and Hassan Bevrani, "Decentralized Model Predictive Control of DC Microgrids with Constant Power Load", *IEEE Journal of Emerging and Selected Topics in Power Electronics*, Vol. 9(1), 2019.

7. A. Srivastava, and R. Bajpai, "An Efficient Maximum Power Extraction Algorithm for Wind Energy Conversion System Using Model Predictive Control", *International Journal on Energy Conversion (IRECON)*, Vol. 7(3), 93–107, 2019.

8. M. Kamran, and Rohail Asghar, "Implementation of Improved Perturb & Observe MPPT Technique with Confined Search Space for Standalone Photovoltaic System", *Journal of king Saud University-Engineering Sciences*, Vol. 32(7), 432–441, 2020.

9. Shazlya Mohamed, and Montaser Abd EI Sattar, "A Comparative Study of P&O and INC Maximum Power Point Tracking Techniques for Grid Connected PV Systems", *SN Applied Sciences*, Vol. 1(2), Article number: 174, 2019, Springer.

10. M. K. Al-Smadi, and Yousef Mohamoud, "Photovoltaic Module Cascaded Converters for Distributed Maximum Power Point Tracking: A Review", *IET Renewable Power Generation*, Vol. 14(4), 2551–2562, 2020.

11. K. Wrobel, P. Serkies, and K. Szabat, "Model Predictive Base Direct Speed Control of Induction Motor Drive-Continuous and Finite Set Approaches", *Energies, MDPI*, Vol. 13(1), 1193, 2020.

12. Sohaib Tahir, J. Wang, M. H. Baloch, and G. S. Kaloi, "Digital Control Techniques based on Voltage Source Inverters in Renewable Energy Applications: A Review", *Application of Power Electronics*, Vol. 7(2), 2018.

13. M. A. Ismeil, A. K. Bakeer, and M. Orabi, "Implementation Quasi Z Source Inverter for PV Applications based on Finite Control Set Model Predictive Control", *International Journal of Renewable Energy Research*, Vol. 9(3), 1462–1471, 2019.

14. M. C. Sha, R. J. Zhu, Xu Guang Yang, and Xu Han, "Finite Control Set Model Predictive Control of Quasi Z-Source Inverter Photovoltaic Grid-Connected", *Journal of Physics: Conference Series*, Vol. 17(48), 2020.

15. Hoach The Nguyen, Jinuk Kim, and Jin Woo Junk, "Improved Model Predictive Control by Robust Prediction and Stability Constraint Finite States for Three Phase Inverter with an Output LC Filter", *IEEE Access*, Vol. 7(1), 12673–12685, 2019.

10 Analysis of an IUPQC Device Using Conventional PID and FOPID Controllers in a Wind Energy Conversion System

G. Pandu Ranga Reddy, Kushal Jagtap, Y. Chintu Sagar, and R. Sheba Rani

10.1 INTRODUCTION

Advancements in power generation, transmission and distribution systems in the recent past have become possible due to improvements in power electronic devices. The usage of power electronics-based loads in the industry has resulted in various power quality issues. These devices also introduce non-sinusoidal harmonic currents to the distribution system, which in turn leads to several adverse effects. Thus, the development of power electronic devices capable of mitigating these effects has gained importance. Many problems are solved with the help of integrating different flexible alternating current transmission (FACT) devices into the system. Static synchronous series compensators (SSSC), series-linked compensators; uniformed power quality conditioners (UPQCs), which combine series and shunt-linked compensators; and static compensators (STATCOMs), which are parallel connected, are the most often-utilised FACTs devices.

A UPQC's inner workings consist of a series active filter coupled to a shunt active filter through a DC link, and vice versa. Due to this, the UPQC will control the current flow through the load and voltage supplied by source. Once this occurs, the voltage and current at the load point are in phase, resulting in sinusoidal voltages and currents throughout all three phases. Due to the high-frequency switching converters, this UPQC has a restricted power flow via the compensator, which is its primary drawback. This can be overcome by designing a novel UPQC, called an improved unified power quality conditioner (IUPQC) device. The block diagram for an IUPQC is similar to a UPQC diagram. Figure 10.1 gives the IUPQC schematic diagram.

The internal control structure of an IUPQC may differ from the internal control structure of a UPQC. The series power converter that is used in a UPQC is powered by a current source that has a sinusoidal waveform, while the voltage source drives

DOI: 10.1201/9781003311829-10

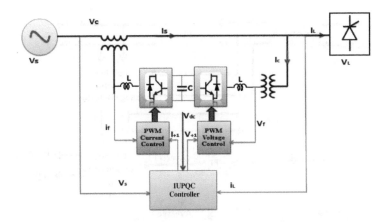

FIGURE 10.1 Improved UPQC block diagram

the shunt converter. To implement an IUPQC, a sinusoidal voltage source drives
the shunt power converter, while a sinusoidal current source drives the series power
converter [1]. Figure 10.2 depicts the internal control system of an IUPQC device. It
can perform the following operations [2, 3].

(i) It can facilitate the transfer of power between the grid and the microgrid.
(ii) It boosts reactive power at the series converter station.
(iii) It is responsible for balancing the frequency and voltage at the shunt converter station.
(iv) In a variety of situations, it can isolate harmonic voltage and current between
series and shunt converter stations.
(v) It is responsible for balancing the voltage and current imbalance.

10.2 IUPQC DEVICE CONTROL STRUCTURE

The IUPQC controller operates on the basis of p-q theory, and its control circuit is
based on the phase locked loop (PLL). At the fundamental positive sequence component, this PLL will deliver a precise frequency and phase angle. The IUPQC tool
works best in a non-neutral, three-phase setup. Therefore, here, the zero sequence
components are not considered. The bus A and B voltages, bus B current (i_L), and
DC link voltage (V_{dc}) serve as controller inputs. Here, the voltages V_{A_ab} and V_{A_bc} are
the source voltages, and I_{L_a} and I_{L_b} are the line currents to the UPQC controller.
First by applying the simplified Clarke transformation, the ABC reference frame is
converted into an α, β reference frame with the following equation [4, 5].

$$\begin{bmatrix} V_{A\_\alpha} \\ V_{A\_\beta} \end{bmatrix} = \begin{bmatrix} 1 & \dfrac{1}{2} \\ 0 & \dfrac{\sqrt{3}}{2} \end{bmatrix} \begin{bmatrix} V_{A\_ab} \\ V_{A\_bc} \end{bmatrix}$$

(10.1)

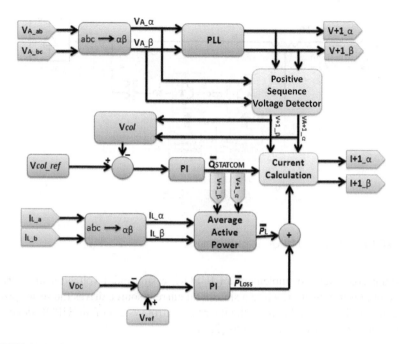

FIGURE 10.2 Internal control structure of an IUPQC

These voltages act as inputs for the PLL. The PLL provides an output signal that has the same phase as the input signal. The PLL circuit uses the PWM technique, and the output signal amplitude is equal to 1 p.u [14]. These voltages may be used as shunt converter reference voltages or as the fundamental +ve sequence component V_{+1_A} of the grid voltage. If the grid voltage amplitude is within the authorised range, V_{+1_A} will prevent the shunt and series converters from drawing unnecessary power [6]. Therefore, the active power is

$$P_L = V_{+1\_\alpha} \cdot i_{L\_\alpha} + V_{+1\_\beta} \cdot i_{L\_\beta} \tag{10.2}$$

Here, i_{L_α} and i_{L_β} are the load currents, and V_{+1_α} and V_{+1_β} are the shunt converter's reference voltages.

The controller's output is the deviation between the setpoint and the grid voltage is [7]

$$V_{cc} = \sqrt{V^2_{A+1\_\alpha} + V^2_{A+1\_\beta}} \tag{10.3}$$

It is possible to determine the reference currents for series converters via

$$\begin{bmatrix} i_{+1\_\alpha} \\ i_{+1\_\beta} \end{bmatrix} = \frac{1}{V^2_{A+1\_\alpha} + V^2_{A+1\_\beta}} \begin{bmatrix} V_{A+1\_\alpha} & V_{A+1\_\alpha} \\ V_{A+1\_\alpha} & V_{A+1\_\alpha} \end{bmatrix} \tag{10.4}$$

10.3 CONVENTIONAL PID CONTROLLER

Because of the simplicity, clear functionality, applicability, and convenience of use provided by proportional–integral–derivative (PID) controllers, most controllers in this world work with traditional PID controllers for industrial applications. If the PID parameters are correctly adjusted, there is no other controller that can match PID controllers for providing robust and dependable performance for most systems. Sometimes the PID controller is also called a three-term control [8]. Figure 10.3 shows a block schematic of a typical PID controller [5].

The PID controller is a kind of feedback mechanism for control loops. The programme calculates the error value by comparing the actual and target values of a process variable. By modifying the process variables, the controller will reduce the error. In this chapter, DC link voltage (V_{dc}), V_{A+1_α}, and V_{A+1_β} are considered process variables. To provide a control action that is optimised for the requirements of a given operation, the PID controller's three constants may be fine-tuned. Many ways of tweaking the PID controller have been devised. In order to construct a process model, the proportional, integral, and derivative parameters are selected in accordance with the model parameters in the majority of the approaches. The Ziegler–Nichols approach is used in this chapter for the purpose of fine tweaking the control settings [9].

10.4 FRACTIONAL ORDER PID CONTROLLER

For a dynamic system, the fractional order (FO) calculus was derived in 1960. Many research articles have reported that the FO controllers can exhibit improved performance compared to conventional controllers. Generally, the fraction order PID controller can be written as "PI$^\lambda$D$^\mu$ controller" [13]. It can provide the extra degree of freedom for integral and derivative terms, along with controller gains (k_p, k_i, k_D). Not all integral and derivative orders follow the same pattern. Figure 10.4 depicts the block diagram of an FOPID controller. It is same as that of a traditional PID controller, but it may allow for more tolerance than a conventional PID controller [10]. An expression for the FOPID controller's transfer function is [11]

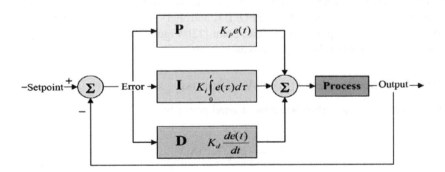

FIGURE 10.3 Conventional PID Controller

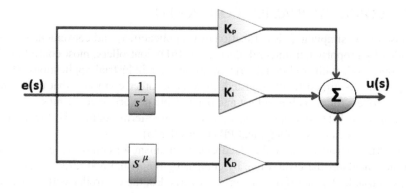

FIGURE 10.4 FOPID controller schematic diagram. On the right-hand side of this figure is an equation that looks like this: e(s). After that, the inverses of sλ and s$^\mu$ are both determined using this equation. Then, there are the values for the parameters K_p, K_i, and K_o. After adding everything together, we get the equation u(s) as our result

$$G_c(s) = k_p + \frac{k_i}{S^\lambda} + K_d S^\mu \tag{10.5}$$

10.5 PROPOSED SYSTEM

This system comprises of two transmission lines that are connected in parallel and connected to the grid, as shown in Figure 10.5. At various time intervals, power quality issues are introduced in the first transmission line, and the second transmission line is considered a testing line [3]. It consists of generating station and distribution station. Here, the generating station is treated as a wind station, and the distribution station consists of an resistive inductive and capacitive (RLC) load. The wind station is primarily made up of three components: a wind turbine (WT), a wind generator (WG) and a converter station. This distribution station is also called the load point. Since the two transmission lines are connected in parallel, the load point on the second line also experiences the aforementioned voltage and current fluctuations due to power quality difficulties. The point of common coupling (PCC) in the second transmission line is where the IUPQC device is linked in order to solve the aforementioned issues. Conventional PID and FOPID controllers are used in an IUPQC device to get the desired response at the load point [12]. Figure 10.5 depicts the complete schematic structure of the proposed power system.

10.5.1 MODELLING EQUATIONS: WIND TURBINE

The equation for the turbine's output of wind power is:

$$P_m = \frac{1}{2}\rho C_p(\lambda,\beta)A_r V_w^3 \tag{10.6}$$

The coefficient of power C_p is calculated by using the formula

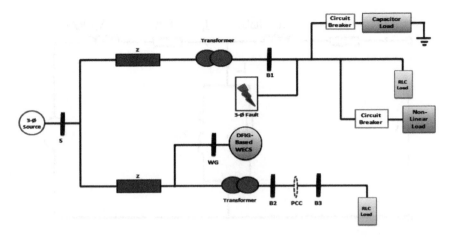

FIGURE 10.5 Proposed system block diagram. An example of a system block diagram has been provided for your reference. The components that make up this system include a source with three phases, an impedance, a transformer, a fault with three phases, a DFIG board WECS, a current breaker, a capacitor load, an RLC load, and a non-linear load

$$C_p(\lambda,\beta) = 0.73\left(\frac{151}{\lambda_i} - 0.58\beta - 0.002\beta^{2.14} - 13.2\right)e^{\frac{18.4}{\lambda_i}}$$

(10.7)

where

$$\lambda_i = \frac{1}{\dfrac{1}{\lambda - 0.02\beta} - \dfrac{0.003}{\beta^3 + 1}}$$

(10.8)

and

$$\text{TSR}\ (\lambda) = \frac{\omega_r R_r}{V_w}$$

(10.9)

10.5.2 Modelling Equations: Wind Generator

In wind energy applications, a variety of generators are offered. Because of its benefits, the double fed induction generator (DFIG) was selected among the others. Advantages of DFIG machines include the ability to independently regulate active and reactive power, the ability to adjust the power factor (PF) by adjusting the voltages applied to the rotors and the ability to acquire the highest possible amount of energy. Figure 10.6 shows the DFIG machine's corresponding circuit.

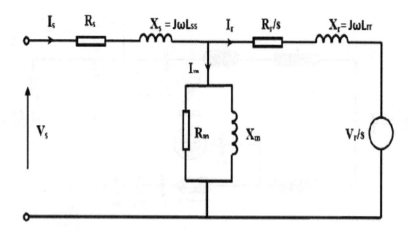

FIGURE 10.6   Equivalent circuit of DFIG

The DFIG machine design equations are derived from the equivalent circuit seen in Figure 10.6 as follows [5].

$$V_{ds} = R_s i_{ds} + \frac{d\psi_{ds}}{dt} - \omega_s \psi_{ds} \tag{10.10}$$

$$V_{qs} = R_s i_{qs} + \frac{d\psi_{qs}}{dt} - \omega_s \psi_{ds} \tag{10.11}$$

$$V_{dr} = R_s i_{dr} + \frac{d\psi_{dr}}{dt} - (\omega_s - \omega_r)\psi_{qr} \tag{10.12}$$

$$V_{qr} = R_s i_{qr} + \frac{d\psi_{qr}}{dt} - (\omega_s - \omega_r)\psi_{dr} \tag{10.13}$$

10.5.2.1   Results Analysis and Discussion

Figure 10.7 shows the MATLAB®/Simulink® proposed system diagram with the IUPQC connected at the PCC. Figure 10.8 shows the internal control structure of the IUPQC. In this system, the voltage sag is established between 0.3 and 0.5 sec, the swell is created between 0.7 and 0.9 and the harmonics are created in between 1.1 and 1.3 sec in the first transmission line. Due to integration, the same effects occur in the second transmission line. This line gives the information about voltage and current waveforms at load point before IUPQC device connected to the system as shown in Figure 10.9 and Figure 10.10.

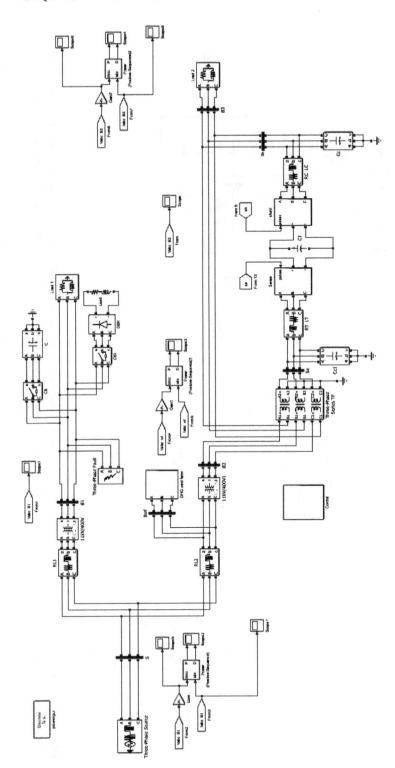

FIGURE 10.7 Simulink diagram of IUPQC device. The IUPQC is a device used to mitigate power quality issues in electrical distribution systems. It combines the functionalities of IPFC and a UPQC to regulate voltage and compensate for disturbances such as harmonics, voltage sags/swells and unbalanced loads. The IUPQC consists of several key components, voltage source converters (VSC), DC link, filters, etc

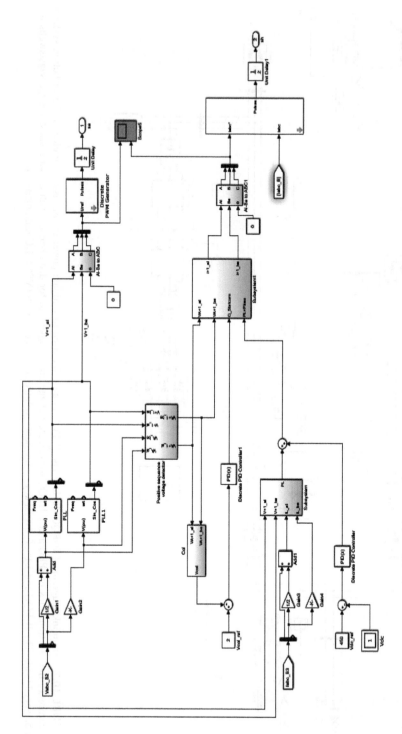

FIGURE 10.8 IUPQC device internal control structure. The internal control structure of an IUPQC typically consists of multiple control loops that regulate the operation of the device. The control structure may vary depending on the specific implementation and control strategy employed

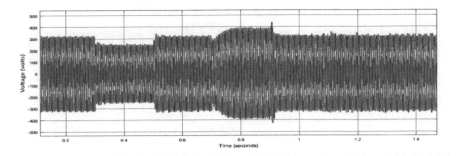

FIGURE 10.9 Voltage at point B3 before the IUPQC is connected

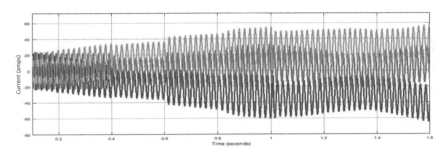

FIGURE 10.10 Current at point B3 before the IUPQC is connected

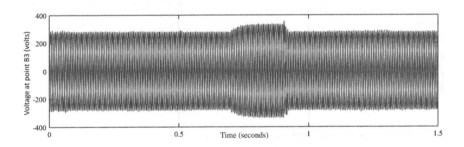

FIGURE 10.11 Voltage at point B3 with a conventional PID controller

10.5.2.2 Simulink Results of an IUPQC Device with a Conventional PID Controller

Voltage and current at load point B3 are shown in Figure 10.11 and Figure 10.12. These graphs show that when a standard PID controller is used to regulate the IUPQC, voltage and current waveform deviations are reduced. Figure 10.13 illustrates the DFIG machine's active and reactive capabilities. The B3 RMS voltage waveform is seen in Figure 10.14. Indicative of voltage sag, a notable drop to 295 V occurs between 0.3 and 0.5 seconds. Additionally, voltage increases to 342 V from 316 V during time $t = 1.1$–1.3 sec owing to voltage swell. Figure 10.15, Figure 10.16 and Figure 10.17 show the percent THD (%THD) of the voltage at point B3 under sag,

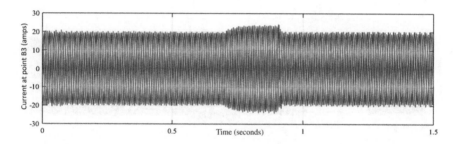

FIGURE 10.12   Current at point B3 with a conventional PID controller

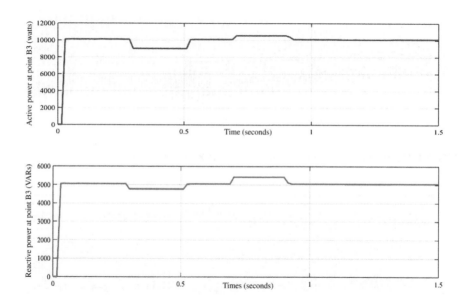

FIGURE 10.13   Active and reactive power at point B3

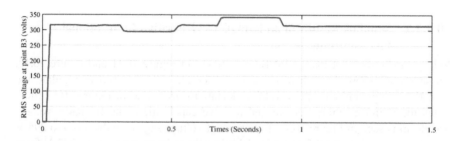

FIGURE 10.14   RMS voltage at point B3

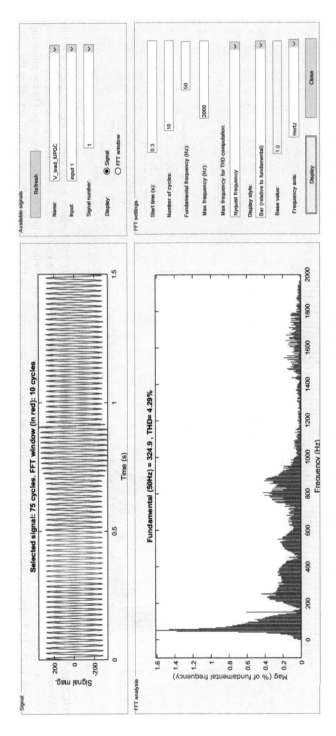

FIGURE 10.15 %THD obtained from the output line voltage at t = 0.3 sec (during sag). The original graph can be seen on the left in this screenshot of the FFT analysis, while the graph produced by the FFT analysis can be seen on the right. The first graph displays time along the x-axis and signal msg along the y-axis. Both of these axes are labelled with their respective units. On the graph that represents the FFT analysis, the frequency is drawn along the x-axis, and the percentage of the fundamental frequency is shown along the y-axis

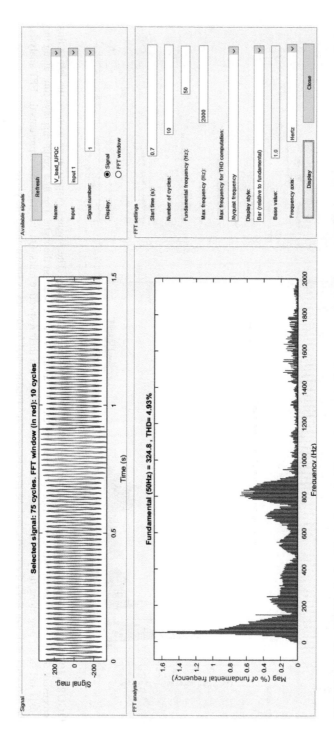

FIGURE 10.16 %THD obtained from the output line voltage at t = 0.7 sec (during swell). The original graph can be seen on the left in this screenshot of the FFT analysis, while the graph produced by the FFT analysis can be seen on the right. The first graph displays time along the x-axis and signal load voltage at t = 0.7 sec with their respective units. On the graph that represents the FFT analysis, the frequency is drawn along the x-axis, and the percentage of the fundamental frequency is shown along the y-axis

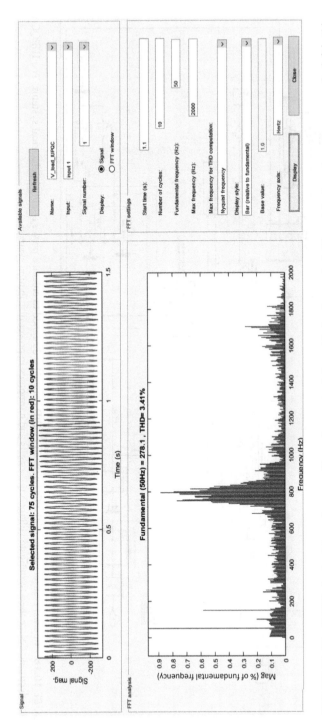

FIGURE 10.17 %THD obtained from the output line voltage at t = 1.1 sec (during harmonics). The original graph can be seen on the left in this screenshot of the FFT analysis, while the graph produced by the FFT analysis can be seen on the right. The first graph displays time along the x-axis and signal msg along the y-axis. Both of these axes are labelled with their respective units. On the graph that represents the FFT analysis, the frequency is drawn along the x-axis, and the percentage of the fundamental frequency is shown along the y-axis

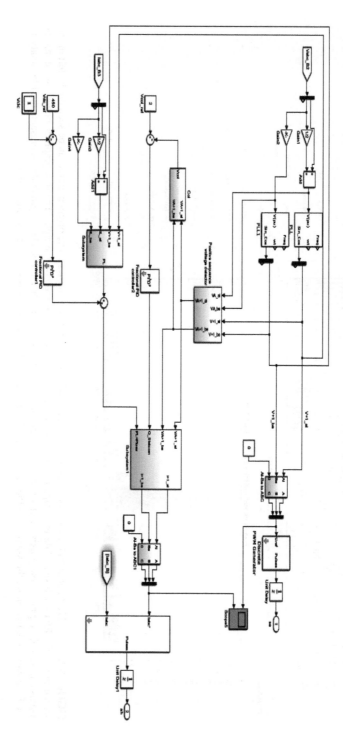

FIGURE 10.18 Internal control circuit of IUPQC with an FOPID controller designed to enhance power quality in electrical systems. The IUPQC is a power electronic device used to mitigate voltage sags, swells and harmonics in power distribution systems

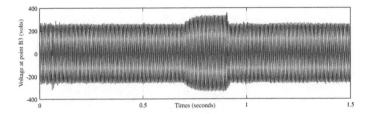

FIGURE 10.19 Voltage at point B3

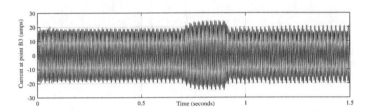

FIGURE 10.20 Current at point B3

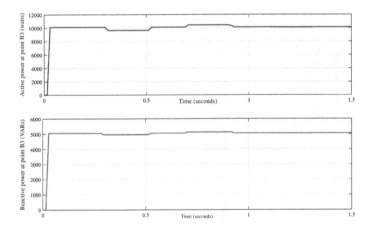

FIGURE 10.21 Active and reactive power at point B3

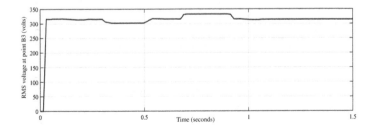

FIGURE 10.22 RMS voltage at point B3

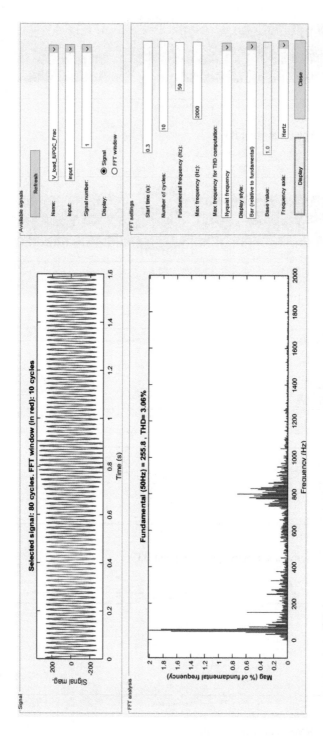

FIGURE 10.23 %THD obtained from the output line voltage at *t* = 0.3 seconds

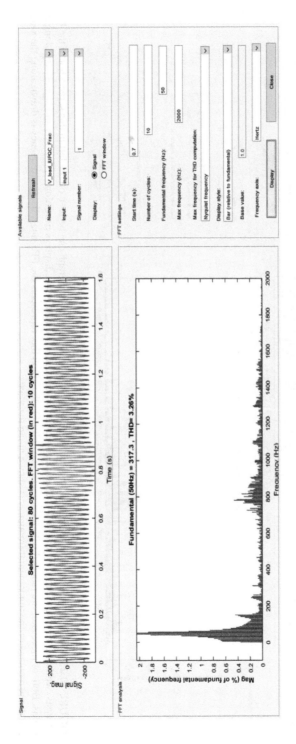

FIGURE 10.24 %THD obtained from the output line voltage at t = 0.7 seconds. The original graph can be seen on the left in this screenshot of the FFT analysis, while the graph produced by the FFT analysis can be seen on the right. The first graph displays time along the x-axis and signal msg along the y-axis. Both of these axes are labelled with their respective units. On the graph that represents the FFT analysis, the frequency is drawn along the x-axis, and the percentage of the fundamental frequency is shown along the y-axis

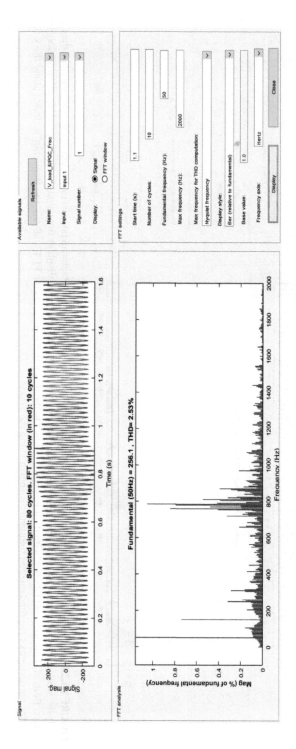

FIGURE 10.25 %THD obtained from the output line voltage at t = 1.1 seconds. The original graph can be seen on the left in this screenshot of the FFT analysis, while the graph produced by the FFT analysis can be seen on the right. The first graph displays time along the x-axis and signal msg along the y-axis. Both of these axes are labelled with their respective units. On the graph that represents the FFT analysis, the frequency is drawn along the x-axis and the percentage of the fundamental frequency is shown along the y-axis

TABLE 10.1

Comparison of proposed controllers with the IUPQC device in terms of %THD

	%THD for V_{load} in the IUPQC	
Time	Conventional PID Controller	FOPID Controller
$t = 0.3$ sec	4.29	3.06
$t = 0.7$ sec	4.93	3.26
$t = 1.1$ sec	3.41	2.53

TABLE 10.2

RMS voltage comparison of proposed controllers using the IUPQC device

	IUPQC RMS Voltage	
Condition	Conventional PID Controller	FOPID Controller
Sag	V_{ref} value 316 V lowered to 295 V	V_{ref} value 316 V lowered to 302 V
Swell	V_{ref} value 316 V raised to 342 V	V_{ref} value 316 V raised to 332 V

swell and harmonic conditions. At $t = 0.3$ sec, the %THD recorded during voltage sag is about 4.29%. The %THD measured at $t = 0.7$ sec during voltage surge is around 4.93%. At a time of $t = 1.1$ sec, %THD is around 3.41% (when harmonics develop owing to non-linear load circumstances).

10.6 SIMULATION RESULTS OF THE IUPQC DEVICE WITH A FOPID CONTROLLER

Figure 10.19 illustrates the voltage and current at point B3 when the IUPQC device is operated with a FOPID controller. Active and reactive power supplied by the DFIG machine at site B3 are shown in Figure 10.21. Figure 10.21 shows that the DFIG machine produces 10 MW of active power and 5 MVar of reactive power. Figure 10.22 depicts the root-mean-square (RMS) voltage waveform collected at point B3. The voltage drops from 316 V to 302 V owing to a sag that happens between $t = 0.3$ and 0.5 sec. A voltage swell causes the voltage to rise from 316 V to 332 V between times $t = 0.7$ and 0.9 sec. The presence of harmonics appears during $t = 1.1$ to 1.3 sec as a result of the connection of non-linear load to the test system. Figure 10.23, Figure 10.24 and Figure 10.25 show the load voltage %THD measured at location B3 at different times for test system operation of the IUPQC device with the FOPID controller. The %THD at $t = 0.3$ sec is 3.06%, at $t = 0.7$ sec is 3.26% and at $t = 1.1$ sec is 2.53%.

10.7 CONCLUSION

In this chapter, different controllers like conventional PID and FOPID controllers were developed for an IUPQC system. Performance of the controllers for the IUPQC device was determined by calculating %THD and RMS voltages at different times. Tables 10.1 and 10.2 show IUPQC %THD and RMS voltages using classic PID and FOPID controllers. From the results, the FOPID controller gives lower %THD and RMS voltage values than the conventional PID controller.

REFERENCES

1. B. W. Franca, and M. Aredes, "Comparisons between the UPQC and Its Dual Topology (iUPQC) in Dynamic Response and Steady-State," in *Proceedings of 37th IEEE IECON*, 2011, pp. 1232–1237.
2. Bruno W. França, Leonardo F. da Silva, Maynara A. Aredes, Maurício Aredes, and Member IEEE, "An Improved iUPQC Controller to Provide Additional Grid-Voltage Regulation as a STATCOM," *IEEE Transactions on Industrial Electronics*, vol. 62, no. 3, March 2015.
3. Dr. G. Pandu Ranga Reddy, Dr. M. Rama Sekhar Reddy, and Dr. M. Vijaya Kumar, "Mitigation of Certain Power Quality Issues in Wind Energy Conversion System Using UPQC and IUPQC Devices," *European Journal of Electrical Engineering*, vol. 22, no. 6, pp. 447–455, December 2020.
4. Ms. Sushma Parihar, Laith O. Maheemed, and Prof. D. S. Bankar, "The Unified Power Quality Conditioner Technique in Wind Energy Conversion System," *International Journal of Advanced Engineering Technology*, pp. 79–82 E-ISSN 0976–3945, April-June 2012.
5. M. Rama Sekhar Reddy, and Dr. M. Vijaya Kumar, "Performance Analysis of Facts Devices for Reduction of Power Quality Issues in DFIGURE Based WECS Integrated to Grid," *International Journal of Engineering & Technology*, vol. 7, pp. 49–56, 2018.
6. J. K. Karanki, G. Geddada, M. K. Mishra, and B. K. Kumar, "A Modified Three-Phase Four-Wire UPQC Topology with Reduced DC-Link Voltage Rating," *IEEE Transactions on Industrial Electronics*, vol. 60, no. 9, pp. 3555–3566, September 2013.
7. Amit Kumar Jindal, Student Member IEEE, Arindam Ghosh, Fellow IEEE, and Avinash Joshi, "Interline Unified Power Quality Conditioner," *IEEE Transactions on Power Delivery*, vol. 22, no. 1, January 2007.
8. J. Sreenivasulu, S. Rajasekhar, A. Pandian, and P. Srinivas Varma, "An Integration of Dual UPQC Controller for Power Quality Compensation by Extending Its Voltage Regulation at Grid Side as a STATCOM," *IOP Conference Series: Materials Science and Engineering*, vol. 993, pp. 1–5, 2020.
9. Sayali Paithankar, and Ranjit Zende, "Comparison between UPQC, iUPQC and Improved iUPQC," in *2017 IEEE 3rd International Conference on Sensing, Signal Processing and Security (ICSSS)*, 2017.
10. Mohamed El-Sayed M. Essa, Magdy A. S. Aboelela, and Mohammed A. M. Hassan, "A Comparative Study between Ordinary and Fractional Order PID Controllers for Modelling and Control of an Industrial system Based on Genetic Algorithm," in *6th International Conference on Modern Circuits and Systems Technologies (MOCAST)*, 2017.
11. M. Aredes, and R. M. Fernandes, "A Dual Topology of Unified Power Quality Conditioner: The iUPQC," in *Proceedings of EPE Conference Applications*, 2009, pp. 1–10.
12. Javvadi. Gowtham Sreeram, and V. Gowtham, "Implementation of Fractional Order PID Controller for An AVR System Using GA Optimization Technique," *International Journal for Research in Applied Science & Engineering Technology (IJRASET)*, vol. 5, no. V, May 2017.
13. M. Al-Dhaifallah, N. Kanagaraj, and K. S. Nisar, "Fuzzy Fractional-Order PID Controller for Fractional Model of Pneumatic Pressure System," *Mathematical Problems in Engineering*, vol.1, pp. 1–9, 2018.
14. G. Kumaraswamy, Y. Rajasekhar Reddy, and C. Harikrishna, "Design of Interline Unified Power Quality Conditioner for Power Quality Disturbances using Simulink," *International Journal of Advancements in Research & Technology*, vol. 1, no. 5, October 2012.

11 Photovoltaic-Based Battery-Integrated E-Rickshaw with Regenerative Braking Using Real-Time Implementation

Arpita Basu and Madhu Singh

11.1 INTRODUCTION

Renewable energy sources have emerged as the most efficient source of energy in the current environment as a result of the rapid depletion of fossil fuels and rising pollution rates. In layman's terms, renewable energy sources are energy sources like small hydro, biomass, wind, and sunlight. When compared to other renewable energy sources, photovoltaic (PV) power generates a lot of power and has become crucial because of its storage and environmental benefits. The PV array will produce more than 45% of the required energy [1]. The PV generation system has two issues, according to Berrera et al., namely low energy conversion efficiency brought on by weather variations and altered solar array power generation [2]. Since there is probably a mismatch between the load characteristics and the maximum power points of PV power generation, maximum power point tracking (MPPT) is crucial. Several researchers, regardless of irradiation and temperature variations, recommended the Perturb & Observe method for MPPT. They demonstrate how utilizing a buck converter can boost the MPPT output [3, 4]. Esram et al. [5] proposed the incremental conductance method (INC), which is the MPPT approach. In the literature, it is explored how different types of MPPT techniques, such as fuzzy control, neural networks, current sweep, etc., compare to one another [6, 7]. A fuzzy logic-based algorithm with INC MPPT approach for PV is proposed by Ali et al. [8]. Since it acts as a conduit between the photovoltaic generation and the modified load, the DC–DC power converter is a crucial part for optimizing a PV solar system. The DC–DC power converter's switch is controlled by the duty cycle signal generated by the MPPT technology. The literature makes use of all three of these converter types: buck, boost, and buck-boost power converters, which are viewed as converters not isolated from the source [9]. A bidirectional converter control technique functions in

DOI: 10.1201/9781003311829-11

the buck–boost and shut-down modes, as suggested by several researchers [10]. The power produced by the PV array determines how the energy storage system (ESS) is charged and discharged [11]. Some publications [12] present the bidirectional converter (BDC) method's traditional control scheme. Due to its limited battery capacity, short range is the main challenge in electric vehicles (EVs). Regenerative braking in EVs has gained more popularity in recent research works. A voltage source inverter (VSI) maintains the regeneration of energy [13, 14]. Regenerative braking ensures that the proper control mechanisms are in place [15]. Regenerative braking is used to recover energy that is wasted when the vehicle brakes. This chapter attempts to analyse an economic approach to solar-powered and battery-powered e-rickshaws along with energy recovery.

In the proposed work, a PV system is mounted on the roof of an e-rickshaw integrated with a lithium-ion battery. This system consists of a 300Watt power PV system, converter circuits, battery storage, and a 1kW brushless direct current (BLDC) motor drive. The PV system, battery storage and BLDC motor drive are connected to a common DC bus. The boost converter uses the incremental conductance (INC) method to implement MPPT utilizing the output of the PV array. The energy storage system, which is coupled to the DC bus, uses a 48V lithium-ion battery. A proportional integral (PI) controller is used to implement closed-loop control of the bidirectional controller, which optimizes the use of the battery management system and controls the flow of electricity for charging and discharging the batteries.

11.2   PV ENERGY SYSTEM

Figure 11.1 depicts the total system's schematic representation. The photovoltaic array, a DC–DC boost converter, a bidirectional DC–DC converter, an energy storage system, and a BLDC drive are the system's five primary components. The PV array's power varies depending on the weather, including temperature and irradiance, and thus it is not always adequate to serve as a source of energy. Hence, a second reliable energy storage system is required. Two converter topologies were proposed by this topology: one is a boost converter, and the other is a DC–DC BDC. By employing a battery management system (BMS), closed-loop control of the BDC is implemented with a proportional integral (PI) controller to regulate the power flow for charging and discharging batteries. The boost converter receives power from the solar system. The MPPT technique is implemented using the INC method. The energy storage system uses a Li-ion battery. The energy flow is controlled by the bidirectional converter during battery charging and draining. The battery starts to deplete when there is less available electricity from the PV system to fulfil the demand. Even with the engine idling, the battery is charged when the power produced by the PV generator is greater than that needed.

11.2.1   SELECTION OF THE PV ARRAY

Due to the limited space (2.6 square metres) in the e-rickshaw, the PV system is calculated with a peak power of 300Watt. A PV array is designed using the solar panel model data sheet in the MATLAB® environment.

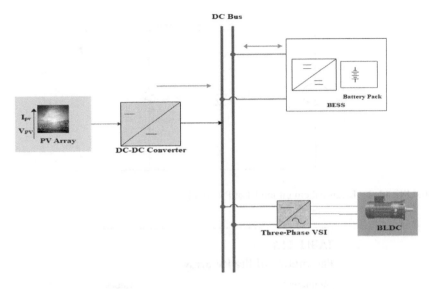

FIGURE 11.1   Schematic diagram of a PV based e-rickshaw

The equivalent circuit of a solar cell is shown in Figure 11.2.

The output current equation of the solar cell can be obtained using equations (11.1) and (11.2) and is given in (11.3).

$$I_{PV} = I_D + I \tag{11.1}$$

$$I_{PV} = I_O \left(e^{\frac{V}{\eta V_T}} - 1 \right) \tag{11.2}$$

Since,

$$I_D = I_O \left(e^{\frac{V}{\eta V_T}} - 1 \right)$$

$$I = I_{PV} = I_O \left(e^{\frac{V}{\eta V_T}} - 1 \right) \tag{11.3}$$

Using the model data sheet in MATLAB®, a 300Watt power PV array is created.

11.2.2.   MPPT OF THE PV ARRAY

The PV array's current and voltage are measured by the MPPT with a pulse width modulation (PWM) control method, which also produces the duty cycle for the converter.

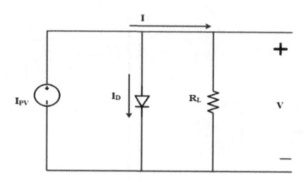

FIGURE 11.2   Electrical circuit model of PV panels

TABLE 11.1
Parameters of the PV array

Parameters	Values
Power	300W
Voc	45.07V
Vmp	25.1V
Isc	8.91A
Imp	8.54A

11.2.2.1   INC Method

INC is derived by differentiating the power of the PV system with the voltage and setting the result equal to zero:

$$\frac{dP_{PV}}{dV_{PV}} = I_{PV}\frac{dV_{PV}}{dV_{PV}} + V_{PV}\frac{dI_{PV}}{dV_{PV}} = I_{PV} + V_{PV} + P_{PV}\frac{dI_{PV}}{dV_{PV}} = 0$$

$$\frac{-I_{PV}}{V_{PV}} = \frac{dI_{PV}}{dV_{PV}} \tag{11.4}$$

Equation (11.4)'s left side denotes the instantaneous conductance's opposite, while its right side denotes the increment. By monitoring the actual values of V_{PV} and I_{PV}, the incremental fluctuations, dV_{PV} and dI_{PV}, can be approximated by both P_{PV} and I_{PV}.

11.2.3   DESIGN OF BOOST CONVERTER

The boost converter's circuit diagram is shown in Figure 11.3. The converter is known as a boost converter because its output voltage exceeds its input voltage.

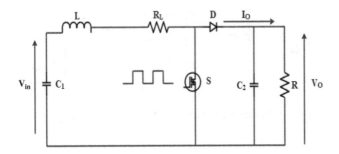

FIGURE 11.3 Schematic diagram of the boost converter

A series-connected boost converter with a PV array is suggested for high efficiency. The output voltage ripple (V) and output current ripple (i_l) are taken to be 5% and 10%, respectively, of the switching frequency (F_{sw}), which is 5 kHz. The boost converter's design specifications are displayed below:

$$\text{Duty Cycle (D)} = 1-(V_{in}/V_{out}) \tag{11.5}$$

V_{in} = input of boost converter = output of PV array.
The inductor value is given as

$$L=V_{pv}D/(2\Delta i_l F_{sw}) \tag{11.6}$$

The output capacitor is given by

$$C=I_o D/(\Delta V F_{SW}) \tag{11.7}$$

11.3 BIDIRECTIONAL DC–DC CONVERTER

Power can be moved from one source to another with the aid of a bidirectional DC–DC converter.

The converter operates in two different modes.

Mode 1 (buck mode: charging mode): The converter controls the charging–discharging process depicted in Figure 11.4 by means of two switches, S_{buck} and S_{boost}. When the buck mode switch (S_{buck}) is turned on, the battery is charged using energy that has been stored in the inductor.

The components that make up the BDC function as a capacitor (C_+) and an inductor (L_+). The converter's values are determined as follows:

$$L_+ = \frac{(V_{dc} - V_{bat})D_+}{\Delta I_L f_{sw}} \tag{11.8}$$

$$C_+ = \frac{(1-D_+)V_{bat}}{8L + \Delta V_{bat} f^2} \tag{11.9}$$

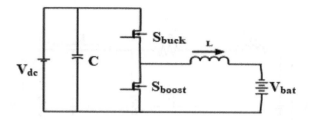

FIGURE 11.4   DC–DC bidirectional converter

where ΔI_L and f_{sw} are the ripple current and switching frequency of the buck converter, respectively.

Mode II (boost mode: discharging mode): The inductor releases the stored energy when the boost mode switch Sboost is on. The load receives the stored energy. the inductor (L_-) and capacitor components (C_-). The converter's values are calculated as follows:

$$L_- = \frac{V_{bat\,D_-}}{\Delta I_L f_{sw}} \tag{11.10}$$

$$C_- = \frac{V_{dc}D_-}{R_O \Delta V_{dc}\, f_{sw}} \tag{11.11}$$

ΔV_{dc} and R_O are the ripple voltage and output voltage of the boost converter, respectively. The values of L and C are considered as follows:

$$L = max(L_+,L_-) \tag{11.12}$$

$$C = max(C_+,C_-) \tag{11.13}$$

11.4   BMS

A PV array's battery bank is connected to the ESS. To prevent overcharging, the bank must be charged. If the battery's overall power exceeds the set limit, the bank will either charge or discharge. The battery's initial state of charge (SOC%) is taken to be 80%. The BMS in this case uses a PI logic controller. After the MATLAB simulation, the controller's improved performance is demonstrated.

Figure 11.5 depicts the dynamic model of a lithium-ion battery, where v(t) is employed to switch between the charging and discharging modes of operation. Battery voltage is determined by:

$$V_{bat} = E_O - K.\frac{S}{S-it}.i*.K.\frac{S}{S-it}.i^* - R.i + A\exp(B-it) \tag{11.14}$$

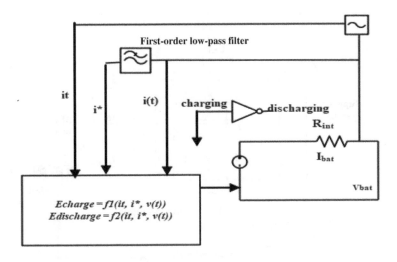

FIGURE 11.5   Lithium-ion battery model

In this equation, V_{bat} is the battery voltage (V), E_O is the constant voltage (V), K is the polarization constant (V/Ah), S is the maximum battery capacity (Ah), A is the exponential voltage (V), B is the exponential capacity (Ah1), R is the internal resistance (ohms), and I is the filtered current. A 48-volt, 200Ah lithium-ion battery that powers the storage system can be charged by the system.

11.5   REGENERATIVE BRAKING METHOD

Figure 11.6 shows the BLDC motor drive with a voltage source inverter. In the proposed method, a back electromotive force (EMF) approach is used to regenerate the drive.

In regenerative mode, the VSI operates like a boost converter. For every 60° interval of electronic commutation, the upper switch halves are turned off and the lower switch halves are in on/off mode. In each step, the PWM control method is used, as in Figure 11.7. After receiving the braking command, the logical switching gate of the inverter is changed. Consequently, regenerative braking is initiated. During the operation mode, all high-side insulated-gate bipolar transistor (IGBT) switches are off, while all low-side IGBT switches are on/off. The whole process is controlled by the PWM method.

The operation of the regenerative mode of the BLDC drive is shown in Figures 11.8(a) and 11.8(b). Here, switches S1, S3, and S5 are turned off. During $0{-}\pi/3$, the lower switch S2, D1, and the winding inductors form the motor driver circuit, which is similar to the boost converter shown in Figure 11.8(b).

Here, 2E = input of the boost converter the output is the DC link or battery.

2L = the inductance of the battery (L = inductance per phase of the system).

The duty ratio of the converter is calculated as

$$\frac{v_b}{2e} = \frac{1}{1-d} \tag{11.15}$$

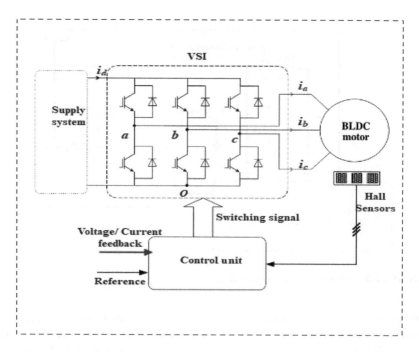

FIGURE 11.6   BLDC motor drive with VSI

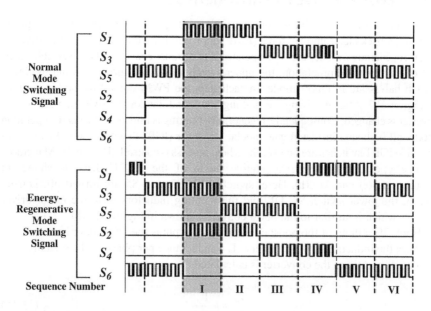

FIGURE 11.7   Switching pulses of VSI during regenerative braking

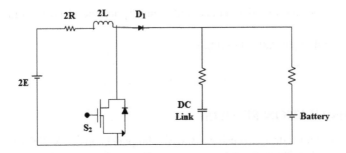

FIGURE 11.8(A) Driver circuit during regenerative braking of BLDC motor

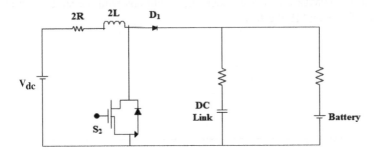

FIGURE 11.8(B) Simplified circuit diagram of BLDC driver circuit

Therefore,

$$d = 1 - \frac{2e}{v_b} \tag{11.16}$$

The current through the motor winding changes because it depends on the load torque. The induction current can be discontinuous under certain operating conditions.

Continuous current flow (CCM) is considered here. At the point where a continuous current flows in the winding, the motor operates at high speed and high torque. The duty cycle (*d*) of the boost converter depends on the battery voltage and the reverse voltage. The speed of the drive controls the back EMF. If the speed is higher, more energy can be recovered from the drive. The KE of the system is given by

$$KE = 0.5mv^2 \tag{11.17}$$

The energy returned to the battery during the regeneration process is

$$E_{waste} = T_L \int_{t_1}^{t_2} \omega(t) \, dt \tag{11.18}$$

Where m is the mass of the vehicle and t_1 and t_2 are the electrical braking time instances.

Hence, total regenerative braking is

$$E_{regenration} = KE - E_{waste} \tag{11.19}$$

11.6   SIMULATION RESULTS

The whole system is designed using a 300W solar panel and a 48V 200Ah Li-ion battery which feeds a 2N-m BLDC motor drive. The system is simulated in MATLAB Environment.

Figure 11.9 shows the speed and electromagnetic torque of the BLDC motor.

Figure 11.10 shows the fundamental operation of the permanent magnet brushless direct current (PMBLDC) drive under idle conditions with the stator current, electromotive force, and Hall effect.

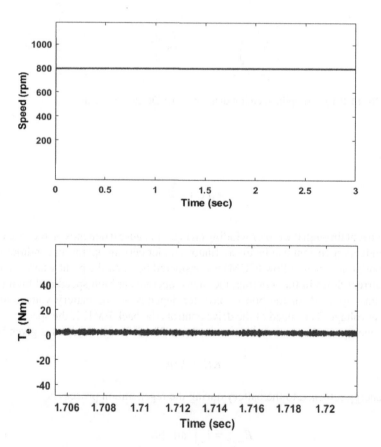

FIGURE 11.9   Speed and torque profiles of the BLDC motor

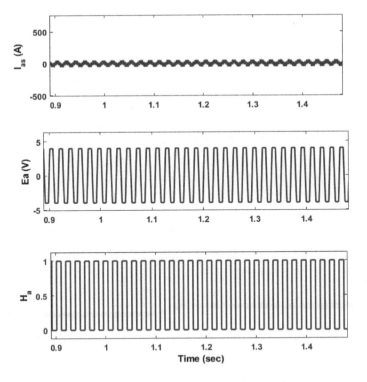

FIGURE 11.10 Stator current, electromotive force, and Hall effect of phase A of the BLDC Motor

Figure 11.11 shows the regenerating operation where the battery starts charging when the motor decelerates and speed reduces. As a result, the electromagnetic torque becomes negative when the SOC increases.

Figure 11.12 shows that at 1.5sec, load torque reduces from 2N-m to 0N-m, the battery SOC level starts charging and the solar irradiation at 1000W/m².

11.7 REAL-TIME DIGITAL SIMULATOR (RTDS) IMPLEMENTATION AND RESULTS DISCUSSION

The networked real-time simulator is a potent modulator that is built utilizing hardware-in-loop and off-line simulation in a MATLAB®/Simulink® model-based design. Figure 11.13 depicts the real-time laboratory (RT-LAB) architecture.

The RT-LAB Simulator architecture is made up of a host and a target. MATLAB and RT-LAB are loaded on the host, a personal computer. The target machine has a Kintex-7 FPGA, 325T, 326,000 logic cells, and 840 DSP slice in addition to an Intel Xeon E3 v5 CPU. TCP/IP is used to connect to the target device and the host computer. Using ethernet, the host PC connects to and communicates with the target. Data observation and real-world simulation verification are performed with a digital storage oscilloscope. These subsystems are named using a prefix that specifies their function in accordance with the RT-LAB naming convention. These are the prefixes' descriptions: 1) the SM main system (always one). It includes the model's

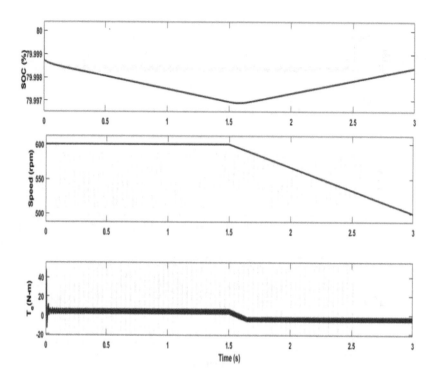

FIGURE 11.11   SOC, speed, and torque in regeneration mode

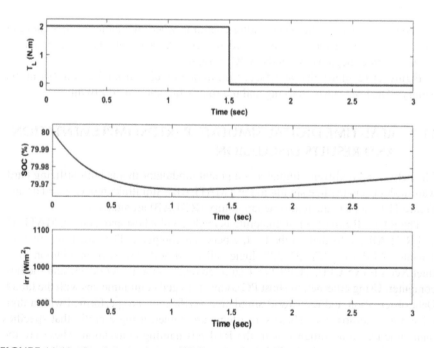

FIGURE 11.12   Load torque, battery SOC, and solar irradiance

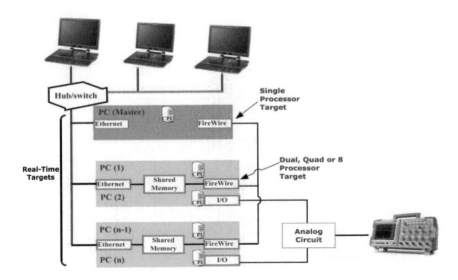

FIGURE 11.13   RT-LAB Simulator architecture

computational components. In this approach, the SM main subsystem is built by assembling the block PV model, boost converter, bidirectional converter, battery, and BLDC motor (Figure 11.14). 2) SC main subsystem (maximum one). In general, all user interface blocks are included. As seen in Figure 11.15, the SC_main subsystem employs the blocks known as scope and manual switches.

The simulation's step size is 0.02 milliseconds. Initially, a PI controller is used in the RT-LAB to implement closed-loop control of the bidirectional converter.

Figure 11.16 details the outcomes of the real-time simulation of the PV system. The PV array's output power at $t = 0–1.5$sec is shown in Figure 11.16(a) to be 300Watt. The solar isolation is 1000W/m^2 at this moment. According to Figure 11.16(b) and (c), at the time period $t = 1.5–3$ sec, the isolation decreases the voltage and current of the PV array both drop while the DC bus voltage, as depicted in Figure 11.16(d), remains constant at about 48V.

The battery is charging when the solar isolation is 1000W/m^2, as shown in Figure 11.17(a). The battery is discharging and the produced power goes down when the solar isolation is 0W/m^2. As shown in Figure 11.17(c), there is reverse flow of the battery current (I_{bat})

Figure 11.18 displays the stator current, motor back EMF voltage, and virtual Hall signal of phase A in real time.

11.8   CONCLUSION

This system consists of a 300W photovoltaic array, converter circuits, battery storage, and a 1kW BLDC motor drive. In this chapter, the study of PV battery management system was introduced with different weather conditions. The DC–DC boost converter is used to regulate the MPPT of PV array. The bidirectional DC–DC converter

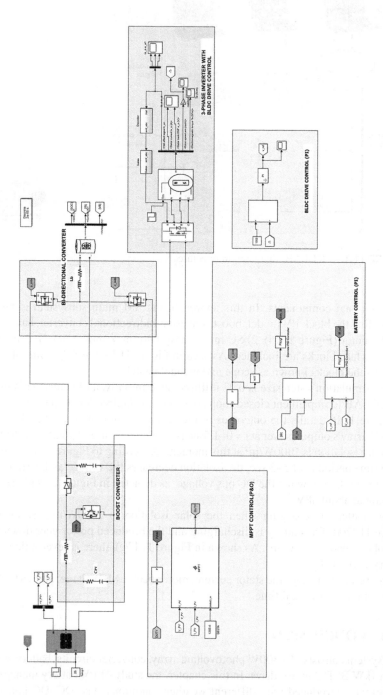

FIGURE 11.14 Configuration of the SM_main subsystem

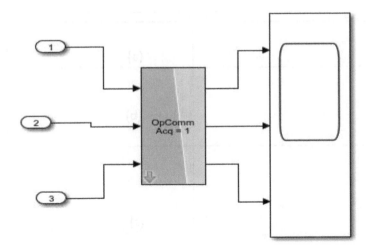

FIGURE 11.15 Configuration of the SC_main subsystem

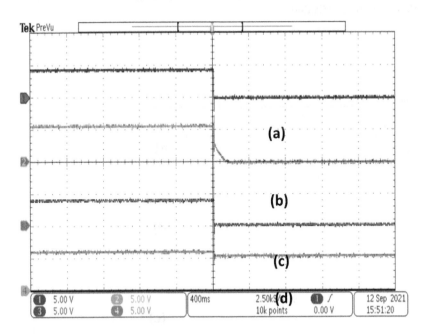

FIGURE 11.16 RTDS hardware results for PV panel (a) P_{PV} (50W/div), (b) V_{PV} (10V/div), (c) I_{PV} (10A/div), and (d) V_{out} (10V/div)

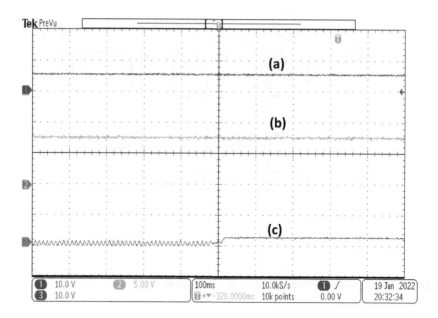

FIGURE 11.17 RTDS hardware results for lithium-ion battery (a) SOC (20/div), (b) V_{bat} (10V/div), and (c) I_{bat} (5A/div)

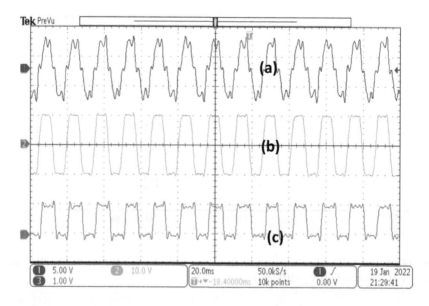

FIGURE 11.18 RTDS hardware results for BLDC motor drive (a) I_s (50V/div), (b) V_{bat} (5V/div), and (c) H_a (5/div)

uses a PI controller scheme. A current control strategy with a PI controller is used. The results show that by changing the value of irradiation, charging or discharging can occur in the battery with respect to the voltage and current. The PV-based battery-integrated e-rickshaw with regenerative braking technology is studied in detail. The regeneration topology has been shown to be satisfactory and could be a solution to the range problem of the lightweight e-rickshaw.

11.9   APPENDIX

Boost converter: output voltage = 48V, duty ratio = 0.17–0.85, inductance = 1.7*mH*, capacitance = 1.8*μF*

DC–DC bidirectional converter: capacitance = 500*μF*, inductance = 1.2*mH*, switching frequency = 5kHz

BLDC: K_E = 0.7rad/sec; poles = 10; stator resistance (R_S) = 0.7ohms; stator inductance (L_S) = 6mH; Kr = 70V/krmp; rated power = 1kW, J = 1.12kgm^2

Battery: nominal voltage = 48V, rated capacity = 200Ah, state-of-charge = 80%. Battery response time = 3sec

REFERENCES

1. Faranda, R., Leva, S., and Maugeri, V. "MPPT techniques for PV systems: Energetic and cost comparison," *IEEE Power and Energy Society General Meeting-Conversion and Delivery of Electrical Energy in the 21st Century*, pp. 1–6, IEEE, July 2008.
2. Berrera, M., Dolara, A., Faranda, R., and Leva, S. "Experimental test of seven widely-adopted MPPT algorithms," *IEEE Bucharest PowerTech*, pp. 1–8, June 2009.
3. Xiao, Weidong, and Dunford, William G. "A modified adaptive hill climbing MPPT method for photovoltaic power systems," *35th Annual Power Electronics Specialists Conference (IEEE Cat. No. 04CH37551)*, vol. 3, pp. 1957–1963, IEEE, 2004.
4. Koutroulis, E., Kalaitzakis, K., and Voulgaris, N. C. "Development of a microcontroller-based, photovoltaic maximum power point tracking control system," *IEEE Transactions on Power Electronics*, vol. 16, no.1, pp. 46–54, 2001.
5. Esram, T., and Chapman, P. L. "Comparison of photovoltaic array maximum power point tracking techniques," *IEEE Transactions on Energy Conversion*, vol. 22, no. 2, pp. 439–449, 2007.
6. Eltamaly, A. M. "Performance of MPPT techniques of photovoltaic systems under normal and partial shading conditions," *Advances in Renewable Energies and Power Technologies*, pp. 115–161, Elsevier, 2018.
7. Kamal, N. A., and Ibrahim, A. M. "Conventional, intelligent, and fractional-order control method for maximum power point tracking of a photovoltaic system: A review," *Fractional Order Systems*, pp. 603–671, Academic Press, 2018.
8. Ali, M. N., Mahmoud, K., Lehtonen, M., and Darwish, M. M. "An efficient fuzzy-logic based variable-step incremental conductance MPPT method for grid-connected PV systems," *IEEE Access*, vol. 9, pp. 26420–26430, 2021.
9. Enrique, J. M., Duran, E., and Sidrach-de-Cardona, M. "Theoretical assessment of the maximum power point tracking efficiency of photovoltaic facilities with different converter topologies," *Solar Energy*, vol. 81, no. 1, pp. 31–38, 2007.
10. Liao, Zhiling, and Ruan, Xinbo. "Control strategy of bi-directional DC/DC converter for a novel stand-alone photovoltaic power system," *IEEE Vehicle Power and Propulsion Conference*, pp. 1–6, IEEE, 2008.

11. Bhattacharyya, S., Puchalapalli, S., and Singh, B. "Battery management and operation of a wind-PV based microgrid," in *IEEE International Conference on Computing, Power and Communication Technologies (GUCON)*, 2020.

12. Sivamani, D. et al. "Solar powered battery charging system using optimized PI controller for buck boost converter," *IOP Conference Series: Materials Science and Engineering*, vol. 1055, no. 1, 2021, IOP Publishing.

13. Hasanah, R. N., Andrean, V., Suyono, H., and Setyawan, R. A. "Bidirectional VSI as a regenerative-braking converter for BLDC motor—An analysis on a plug-in electric vehicle application," in *10th International Conference on Electrical and Electronics Engineering (ELECO)*, pp. 222–226, 2017.

14. Mishra, S., Varshney, A., Singh, B., and Parveen, H. "Driving cycle based modelling and control of solar-battery fed RelSyn motor drive for light electric vehicle with energy regeneration," in *2021 IEEE Energy Conversion Congress and Exposition (ECCE)*, pp. 5008–5013, 2021.

15. Deepa, M. U., and Bindu, G. R. "A novel switching scheme for regenerative braking and battery charging for BLDC motor drive used in electric vehicle," in *IEEE International Power and Renewable Energy Conference*, pp. 1–6, 2020.

12 Allocation of Distribution System Losses Considering the Effects of Load Power Factor and Distributed Generation

Kushal Jagtap, Vijay Pal Singh, and Aijaz Ahmad

12.1 INTRODUCTION

In a restructured power system, transmission and distribution systems are still vertically integrated systems with natural monopolies, and network operations expenses must be distributed to users through network usage tariffs that consider their actual impact on prices. One of these expenses is the cost of distribution power losses, which is challenging to allocate due to the nonlinear relationship between losses and injected power. It is difficult to determine how each user's consumption affects network losses and how these losses should be distributed among network users, including consumers and distributed generators. The authors of a study [1] concluded that regardless of the method used, a certain degree of arbitrariness always exists in the allocation of losses due to their non-separable and nonlinear nature. It is impossible to separate losses into terms that are individually attributable to load or generation.

To be effective, a method for allocating network losses should be based on actual network data and be easily comprehensible. It should also be economically efficient, avoiding cross-subsidization among users while conveying economic signals that promote overall network efficiency. Furthermore, the method must recover all system losses, be unbiased towards consumers, be consistent across various scenarios, and be applicable to electrical markets that have open access.

The allocation of losses has traditionally been a concern for transmission networks, but it has become increasingly important for distribution systems as well due to the rising use of distributed generation and the introduction of supplier competition. Distribution systems were originally designed for unidirectional power flow, but the integration of distributed generation has presented new challenges. While distributed generation can offer several benefits such as voltage support, reduced power loss, support for auxiliary services, and improved reliability [2, 3], it can also have a range of impacts on the operation of distribution systems.

In addition, distribution systems cater to various sectors such as residential, industrial, and commercial, with consumers classified as low or medium voltage users. The operational performance of these consumers is impacted by fluctuations in their power factor (PF). The loading of distribution lines is affected by changes in PF, as the PF of the system influences the quality of the load.

The distribution system operates in a radial configuration but experiences losses due to the high R/X ratio of distribution lines. Fluctuations in the PF of consumers and local power injection by distributed generators (DGs) also impact the loading of distribution lines [4]. Therefore, it is crucial to determine the contributions of consumers and DGs to system losses by tracing their active and reactive power flow [5] to ensure that no consumers or DGs are unfairly burdened with the cost of system losses.

The literature provides various methods for allocating system losses in distribution systems. Nikolaidis et al. [5] developed a graph-based mechanism that links the physical characteristics of an imbalanced radial distribution system with financial activities in energy markets. Usman et al. [6] proposed a multi-phase loss-allocation system that explicitly assigned losses associated with phase-current cross-terms to the neutral. Moret et al. [7] accounted for grid restrictions and line losses by treating system operators at both transmission and distribution levels as active market participants. Pankaj Kumar et al. [8] discussed a circuit-based branch current decomposition approach for loss allocation in an active modified distribution system with reactive power transactions and superposition theory for rewards/penalties for DG owners. Shafeeque Ahmed and Karthikeyan [9] employed the load sensitivity factor and loss contribution fraction of multiple transactions to evenly divide the real and reactive loss components among the loads. Zarabadipoura and Mahmoudib [10] used a weighted sharing and voltage sensitivity approach to allocate system losses based on power injections and Z-bus matrices. Moon et al. [11] developed a path-integral approach for assigning bus-wise transmission loss using the loss sensitivity market-operating strategy.

Khosravi et al. [12] developed an algorithm based on line active and reactive power flows, which allowed for the identification of line losses caused by the load and the supplier separately. Alayande et al. [13] presented a solution based on inherent structural characteristics theory for allocating losses to network participants and addressing generation-to-load matching problems. Hota and Mishra [14] proposed an active power loss-allocation approach using forward–backward sweep power flow, with or without distributed generators. Yu et al. [15] proposed a modified Shapley value method with stratified sampling to reduce the number of samples collected. To overcome the problem of combinational explosion in traditional cooperative-game-based allocation methods, Yu et al. [16] proposed a minimum cost-remaining savings method and the Aumann–Shapley value method for loss and emission reduction allocation. Amaris et al. [17] combined electrical circuit theory with Aumann–Shapley game theory to calculate a unitary participation coefficient for each network user using network structure and currents demanded/injected. Al-Digs and Chen [18] focused on the loss divider problem and provided an analytical closed-form equation for nodal active and reactive power injections in a steady-state distribution system.

The existing literature lacks studies that demonstrate how system losses are allocated when line loading fluctuates due to changes in load PF and the installation of DGs. Therefore, this paper proposes a mechanism for assigning distribution system losses that considers the dynamic changes in line loading. The chapter is divided into three sections: 1) the proposed method addresses the changes in line loading due to load PF variations; 2) the proposed method determines system losses and assigns them exclusively to consumers using a natural decomposition technique; and 3) the proposed method assigns maximum benefits to distributed generations for their contribution to system performance improvement by extending the decomposition technique. The proposed method is tested on a 28-bus radial distribution system, and the results are compared with those from proportional and quadratic algorithms [19].

The rest of the chapter is divided into three sections. The second section outlines the mathematical formulae for determining and assigning system losses to consumers using a natural decomposition technique. Also, the section extends the decomposition technique to assign maximum benefits to distributed generations for their contribution to system performance. The third section discusses the results obtained by the proposed method and compared with that of other methods. Finally, section 12.4 concludes the work.

12.2   MATHEMATICAL FORMULATION

The goal of this chapter is accomplished through three distinct subsections, each with its own mathematical modeling, which is developed in this section. In subsection 12.2.1, the change in power due to the alteration in PF is calculated. In subsection 12.2.2, the network losses caused by consumers are calculated. Finally, in subsection 12.2.3, the allocation of loss savings to DGs is computed.

12.2.1   CHANGE IN POWER DUE TO VARIATIONS IN POWER FACTOR

The system PF defines the quality of the load. Distribution system operators require consumers to operate at specific PFs to improve system performance and efficiency. However, the PF of consumers may deviate from the specified value due to the operating conditions of their inductive loads. To evaluate this, the chapter considers three different PF scenarios: when the consumer PF is equal to the specified/reference PF, when it is less than the specified PF, and when it is greater than the specified PF.

The method used in this section to calculate the change in power resulting from PF variations is based on a geometrical approach. In this approach, a circle with a constant radius representing the apparent power is used for each PF angle, as illustrated in Figure 12.1.

From Figure 12.1, apparent power of load connected at bus i, i.e., load i, on the reference PF angle ϕ_{rf} is stated as follows:

$$S_{rf}\{l,i\} = \sqrt{P_{rf}\{l,i\}^2 + Q_{rf}\{l,i\}^2} = \frac{P_{rf}\{l,i\}}{\cos\phi_{rf}} \tag{12.1}$$

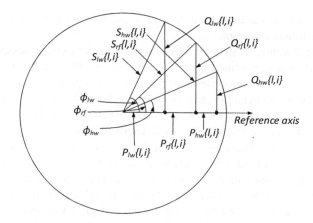

FIGURE 12.1   Circle diagram

where, $\cos\phi_{rf}$ is the reference PF, and $S_{rf}\{l,i\}$, $P_{rf}\{l,i\}$, and $Q_{rf}\{l,i\}$ are the apparent, active, and reactive power of load i on the reference PF.

Similarly, the apparent power of load i on the low and the high PF angles ϕ_{lw} and ϕ_{hw}, respectively, is expressed as:

$$S_{lw}\{l,i\} = \frac{P_{lw}\{l,i\}}{\cos\phi_{lw}} = \frac{P_{rf}\{l,i\} + \Delta P_{lw}\{l,i\}}{\cos\phi_{lw}} \tag{12.2}$$

and

$$S_{hw}\{l,i\} = \frac{P_{hw}\{l,i\}}{\cos\phi_{hw}} = \frac{P_{rf}\{l,i\} + \Delta P_{hw}\{l,i\}}{\cos\phi_{hw}} \tag{12.3}$$

where $\cos\phi_{lw}$ and $\cos\phi_{hw}$ are low and high PF, respectively; $S_{lw}\{l,i\}$ and $P_{lw}\{l,i\}$ are the apparent and active power of load i on the low PF, respectively; $S_{hw}\{l,i\}$ and $P_{hw}\{l,i\}$ are the apparent and active power of load i on the high PF, respectively; and $\Delta P_{lw}\{l,i\}$ and $\Delta P_{hw}\{l,i\}$ are change in active power on low and high PF, respectively.

According to Figure 12.1, the hypotenuse of the power triangle remains constant at each instant of the PF angles, satisfying the following condition:

apparent power at ϕ_{rf} = apparent power at ϕ_{lw}

Equation (12.4) will be obtained by equating Eq. (12.1) and Eq. (12.2), thus:

$$\frac{P_{rf}\{l,i\}}{\cos\phi_{rf}} = \frac{P_{rf}\{l,i\} + \Delta P_{lw}\{l,i\}}{\cos\phi_{lw}} \tag{12.4}$$

After rearranging Eq. (12.4), the expression for $\Delta P_{lw}\{l,i\}$ is given by:

$$\Delta P_{lw}\{l,i\} = \frac{\cos\phi_{lw} - \cos\phi_{rf}}{\cos\phi_{rf}} \times P_{rf}\{l,i\} \tag{12.5}$$

Similarly, rearranging after equating Eq. (12.1) and Eq. (12.3), the expression for $\Delta P_{hw}\{l,i\}$ is given by:

$$\Delta P_{hw}\{l,i\} = \frac{\cos\phi_{hw} - \cos\phi_{rf}}{\cos\phi_{rf}} \times P_{rf}\{l,i\} \qquad (12.6)$$

Active power drawn by the load on low PF, using Eq. (12.2) and Eq. (12.5), is given by:

$$P_{lw}\{l,i\} = P_{rf}\{l,i\} + \Delta P_{lw}\{l,i\} = \left(1 + \frac{\cos\phi_{lw} - \cos\phi_{rf}}{\cos\phi_{rf}}\right) \times P_{rf}\{l,i\} \qquad (12.7)$$

$$P_{lw}\{l,i\} = CoALW \times P_{rf}\{l,i\} \qquad (12.8)$$

where $CoALW = \left(1 + \dfrac{\cos\phi_{lw} - \cos\phi_{rf}}{\cos\phi_{rf}}\right)$ is the coefficient of low PF for the active load.

Active power drawn by the load with high PF, using Eq. (12.3) and Eq. (12.6), is given by:

$$P_{hw}\{l,i\} = P_{rf}\{l,i\} + \Delta P_{hw}\{l,i\} = \left(1 + \frac{\cos\phi_{hw} - \cos\phi_{rf}}{\cos\phi_{rf}}\right) \times P_{rf}\{l,i\} \qquad (12.9)$$

$$P_{hw}\{l,i\} = CoAHW \times P_{rf}\{l,i\} \qquad (12.10)$$

where $CoAHW = \left(1 + \dfrac{\cos\phi_{hw} - \cos\phi_{rf}}{\cos\phi_{rf}}\right)$ is the coefficient of high PF for the active load.

Further, from Figure 12.1, the apparent power of load i on the reference PF angle ϕ_{rf} can be expressed as:

$$S_{rf}\{l,i\} = \frac{Q_{rf}\{l,i\}}{\sin\phi_{rf}} \qquad (12.11)$$

where $Q_{rf}\{l,i\}$ is the reactive power of load i on the reference PF.

Similarly, the apparent power of load i on the low and the high PF angles ϕ_{lw} and ϕ_{hw}, respectively, is expressed as:

$$S_{lw}\{l,i\} = \frac{Q_{lw}\{l,i\}}{\sin\phi_{lw}} = \frac{Q_{rf}\{l,i\} + \Delta Q_{lw}\{l,i\}}{\sin\phi_{lw}} \qquad (12.12)$$

and
$$S_{hw}\{l,i\} = \frac{Q_{hw}\{l,i\}}{\sin\phi_{hw}} = \frac{Q_{rf}\{l,i\} + \Delta Q_{hw}\{l,i\}}{\sin\phi_{hw}} \qquad (12.13)$$

The condition, apparent power at ϕ_{rf} = apparent power at ϕ_{hw}, is valid, and Eq. (12.14) can be obtained by equating Eq. (12.11) and Eq. (12.12), thus:

$$\frac{Q_{rf}\{l,i\}}{\sin\phi_{rf}} = \frac{Q_{rf}\{l,i\} + \Delta Q_{hw}\{l,i\}}{\sin\phi_{hw}} \qquad (12.14)$$

After rearranging Eq. (12.14), the expression for $\Delta Q_{lw}\{l,i\}$ is given by:

$$\Delta Q_{lw}\{l,i\} = \frac{\sin\phi_{lw} - \sin\phi_{rf}}{\sin\phi_{rf}} \times Q_{rf}\{l,i\} \qquad (12.15)$$

Similarly, rearranging after equating Eq. (12.11) and Eq. (12.13), the expression for $\Delta Q_{hw}\{l,i\}$ is given by:

$$\Delta Q_{hw}\{l,i\} = \frac{\sin\phi_{hw} - \sin\phi_{rf}}{\sin\phi_{rf}} \times Q_{rf}\{l,i\} \qquad (12.16)$$

Reactive power drawn by the load on low PF, using Eq. (12.12) and Eq. (12.15), is given by:

$$Q_{lw}\{l,i\} = Q_{rf}\{l,i\} + \Delta Q_{lw}\{l,i\} = \left(1 + \frac{\sin\phi_{lw} - \sin\phi_{rf}}{\sin\phi_{rf}}\right) \times Q_{rf}\{l,i\} \qquad (12.17)$$

$$Q_{lw}\{l,i\} = CoRLW \times Q_{rf}\{l,i\} \qquad (12.18)$$

where $CoRLW = \left(1 + \dfrac{\sin\phi_{lw} - \sin\phi_{rf}}{\sin\phi_{rf}}\right)$ is the coefficient of low PF for the reactive load.

Reactive power drawn by the load on high PF, using Eq. (12.13) and Eq. (12.16), is given by:

$$Q_{hw}\{l,i\} = Q_{rf}\{l,i\} + \Delta Q_{hw}\{l,i\} = \left(1 + \frac{\sin\phi_{hw} - \sin\phi_{rf}}{\sin\phi_{rf}}\right) \times Q_{rf}\{l,i\} \qquad (12.19)$$

$$Q_{hw}\{l,i\} = CoRHW \times Q_{rf}\{l,i\} \qquad (12.20)$$

where, $CoRHW = \left(1 + \dfrac{\sin\phi_{hw} - \sin\phi_{rf}}{\sin\phi_{rf}}\right)$ is the coefficient of high PF for the reactive load.

12.2.2 ALLOCATION OF SYSTEM LOSSES

In this section, losses are calculated by removing DGs (represented by a dotted line) from the distribution network shown in Figure 12.2. Branch power flows upward from bus i to bus $i+1$, and downstream branches draw additional power from bus $i+1$. Before implementing the proposed mechanism, an AC power flow analysis is necessary to determine the voltage levels at each node.

Assume the intermediate branch i of the n-bus distribution system is connected between its sending end bus i and receiving end bus $i+1$, as shown in Figure 12.2. By avoiding the effect of the shunt element, power at the receiving end of branch i is a function of power drawn by the load connected at bus $i+1$, i.e., load $i+1$ and power flow in the downward branches omitted from bus $i+1$. By applying Kirchhoff's law, receiving end power can be expressed as:

$$P\{i\} + jQ\{i\} = \left(P\{l,i+1\} + jQ\{l,i+1\}\right) + \sum_{k \in Br_i}\left(P'\{k\} + jQ'\{k\}\right) \quad (12.21)$$

where, $P\{i\}$ and $Q\{i\}$ are the receiving end active and reactive power, respectively, of branch i; $P'\{i\}$ and $Q'\{i\}$ are the sending end active and reactive power, respectively, of branch i; $P\{l,i+1\}$ and $Q\{l,i+1\}$ are the active and reactive power, respectively, of load $i+1$; and Br_i is the set of branches connected downward of branch i.

Power at the sending end of branch i can be expressed as

$$P'\{i\} + jQ'\{i\} = \left(P\{ls,i\} + jQ\{ls,i\}\right) + \left(P\{i\} + jQ\{i\}\right) \quad (12.22)$$

where $P\{ls,i\}$ and $Q\{ls,i\}$ are the active and reactive power loss, respectively, of branch i.

Further, complex power loss in the branch i is given by:

$$P\{ls,i\} + jQ\{ls,i\} = \frac{R\{i\} + jX\{i\}}{\left|V\{i+1\}\right|^2} \times \left(P\{i\}^2 + Q\{i\}^2\right) \quad (12.23)$$

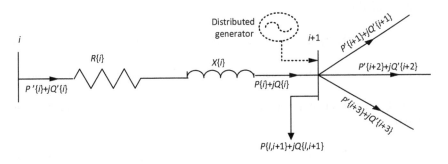

FIGURE 12.2 Sample branch of an n-bus distribution system

where, $R\{i\}$ and $X\{i\}$ are the resistance and reactance, respectively, of branch I, and $V\{i+1\}$ is the phasor voltage of the bus $i+1$.

After decomposing injected power in Eq. (12.23), active power loss in the branch i is given by:

$$P\{ls,i\} = \frac{R\{i\} \times P\{i\}}{|V\{i+1\}|^2} \times P\{i\} + \frac{R\{i\} \times Q\{i\}}{|V\{i+1\}|^2} \times Q\{i\} \tag{12.24}$$

$$P\{ls,i\} = \alpha_i \times P\{i\} + \beta_i \times Q\{i\} \tag{12.25}$$

where $\alpha_i = \dfrac{R\{i\} \times P\{i\}}{|V\{i+1\}|^2}$ is the coefficient of active power injection belonging to active loss and

$\beta_i = \dfrac{R\{i\} \times Q\{i\}}{|V\{i+1\}|^2}$ is the coefficient of reactive power injection belonging to active loss.

Using Eqs. (12.21), (12.22), and (12.25), the active power loss of branch i can also be written as:

$$P\{ls,i\} = \alpha_i \times \left(P\{l,i\} + \sum_{k \in Br_i} \left(P\{ls,k\} + P\{k\} \right) \right) + \beta_i \times \left(Q\{l,i\} + \sum_{k \in Br_i} \left(Q\{ls,k\} + Q\{k\} \right) \right) \tag{12.26}$$

But $\alpha_i \times P\{ls,k\} <<< \alpha_i \times P\{l,k\}$; therefore, to simplify the complexity of the allocation mechanism generally, the terms $\alpha_i \times P\{ls,k\}$ are eliminated [20]. The same holds true for reactive power terms.

By starting the computation and allocation of losses from the terminal branch, the need for an additional reconciliation step can be avoided. Allowing such a practice to calculate system losses may result in an error, which is the difference between the algebraic sum of allocated losses and the system losses obtained using the power flow algorithm.

Now, active power loss in the terminal branch $t-1$ allocated to the consumer t is calculated as follows:

$$\Pi P\{ls,t-1,l,t\} = \alpha_{t-1} \times P\{l,t\} + \beta_{t-1} \times Q\{l,t\} \tag{12.27}$$

where $\Pi P\{ls,t-1,l,t\}$ is the allocated active power loss in the branch $t-1$ to load on bus t.

Similarly, for reactive power loss allocation in the terminal branch $t-1$ is expressed as:

$$\Pi Q\{ls,t-1,l,t\} = \alpha_{t-1} \times P\{l,t\} + \beta_{t-1} \times Q\{l,t\} \tag{12.28}$$

where $\prod Q\{ls, t-1, l, t\}$ is the allocated reactive power loss in the branch t-1 to load on bus t,

$\alpha_{t-1} = \dfrac{X\{t-1\} \times P\{t-1\}}{|V\{t\}|^2}$ is the coefficient of active power injection belonging to

reactive loss, and

$\beta_{t-1} = \dfrac{X\{t-1\} \times Q\{t-1\}}{|V\{t\}|^2}$ is the coefficient of reactive power injection belonging to reactive loss.

After allocating power loss in terminal branch t-1, the active and reactive power of the load t is updated as follows:

$$P\{l, t, t-1\} = P\{l, t\} + P\{ls, t-1, l, t\} \tag{12.29}$$

$$Q\{l, t, t-1\} = Q\{l, t\} + Q\{ls, t-1, l, t\} \tag{12.30}$$

where $P\{l, t, t-1\}$ and $Q\{l, t, t-1\}$ are the updated active and reactive power of the load on bus t, respectively, due to the allocated active power loss of branch t-1.

After updating the load power, the power loss in the preceding branch t-2 is expressed as:

$$\prod P\{ls, t-2, l, t\} = \alpha_{t-2} \times P\{l, t, t-1\} + \beta_{t-2} \times Q\{l, t, t-1\} \tag{12.31}$$

$$\prod P\{ls, t-2, l, t-1\} = \alpha_{t-2} \times P\{l, t-1\} + \beta_{t-2} \times Q\{l, t-1\} \tag{12.32}$$

Similarly, expressions for reactive loss allocation are:

$$\prod Q\{ls, t-2, l, t\} = \alpha_{t-2} \times P\{l, t, t-1\} + \beta_{t-2} \times Q\{l, t, t-1\} \tag{12.33}$$

$$\prod Q\{ls, t-2, l, t-1\} = \alpha_{t-2} \times P\{l, t-1\} + \beta_{t-2} \times Q\{l, t-1\} \tag{12.34}$$

The procedure for allocating losses is continued until all branches are tallied or the reference bus is reached. After accounting for all branches, the final value of loss allocation is derived by subtracting the rated power of load i from its updated values, as follows:

$$\prod P\{ls, l, i\} = P\{l, i, n-1\} - P\{l, i\} \tag{12.35}$$

where $\prod P\{ls, l, i\}$ is the allocated active power loss to load i, and n is the total number of buses in the system.

Finally, to ensure the efficiency of the proposed method, total system losses must equal the sum of allocated losses, which can be represented as:

$$P\{ls\} = \sum_{i=1}^{n} \left(\Pi P\{ls,l,i\} \right) \tag{12.36}$$

where $P\{ls\}$ is the active power loss of distribution systems as determined by the power flow algorithm.

12.2.3 ALLOCATION OF LOSS SAVING

In this subsection, the focus is on providing maximum operational benefits to DGs. The distribution network can be modified by incorporating DGs in optimal locations with suitable sizes, as illustrated in Figure 12.2. The loss variations caused by the DGs are analyzed and assigned to them. When DGs inject power into the network, it alters the power flow and network losses compared to the base case (without DGs). DGs contribute to network loss reduction when they are strategically located and scaled within the network. Therefore, a new power flow is executed in the presence of DGs, and the active and reactive power flows in each branch are calculated.

Power reached at the receiving end of branch i can be expressed by modifying Eq. 12.21, as follows:

$$P_g\{i\} + jQ_g\{i\} = \left(\left(P\{l,i+1\} - P\{g,i+1\} \right) + j\left(Q\{l,i+1\} - Q\{g,i+1\} \right) \right) + \sum_{k \in Br_i} \left(\left(P_g\{k\} + jQ_g\{k\} \right) + \left(P_g\{ls,k\} + jQ_g\{ls,k\} \right) \right) \tag{12.37}$$

where $P_g\{i\}$ and $Q_g\{i\}$ are the receiving end active and reactive power of branch i, respectively, in the presence of DGs; $P\{g,i\}$ and $Q\{g,i\}$ are the active and reactive power of DGs, respectively, on bus i; and $P_g\{ls,i\}$ and $Q_g\{ls,i\}$ are the active and reactive power loss of branch i, respectively, in the presence of DGs.

In this section, it is assumed that the node voltages in the presence of distributed generators are the same as those in the base case. This is because the operation of distribution systems is sensitive to changes in node voltages, and corrective measures are taken promptly if any deviations from the rated voltage magnitude occur.

Active power loss of branch i in the presence of DGs is expressed as:

$$P_g\{ls,i\} = A_i \times \sum_{k \in Nd_i} \left(P\{l,k\} - P\{g,k\} + P_g\{ls,k+1\} \right) + B_i \times \sum_{k \in Nd_i} \left(Q\{l,k\} - Q\{g,k\} + Q_g\{ls,k+1\} \right) \tag{12.38}$$

where $P_g\{ls,i\}$ is the active power loss of branch i in the presence of DGs; Nd_i is the set of nodes ahead of branch i; $A_i = \dfrac{R\{i\} \times P_g\{i\}}{|V\{i+1\}|^2}$ is the coefficient of active power injection in the presence of DGs belonging to active loss; and $B_i = \dfrac{R\{i\} \times Q_g\{i\}}{|V\{i+1\}|^2}$ is

the coefficient of reactive power injection in the presence of DGs belonging to active loss.

The active power loss savings of the network are represented as a function of losses without and with the inclusion of DGs, defined as:

$$P_{sav}\{ls\} = P\{ls\} - P_g\{ls\} \qquad (12.39)$$

where $P_{sav}\{ls\}$ is the active power loss saving of the system, and $P_g\{ls\}$ is the active power loss of the system in the presence of DGs.

Complex power flow reduction in the branch i due to distributed generators in the network is:

$$P_{red}\{i\} + jQ_{red}\{i\} = \left(P\{i\} + jQ\{i\}\right) - \sum_{k \in Nd_i} \left(P\{g,k\} + jQ\{g,k\}\right) \qquad (12.40)$$

where $P_{red}\{i\}$ and $Q_{red}\{i\}$ are the reductions in the active and reactive power flow of branch i, respectively.

The contribution of DGs to bus i for system loss saving is expressed as:

$$\Pi P_{sav}\{ls,g,i\} = P\{ls\} - \sum_{k \in \mu_i} \left(A_k \times P_{red}\{k\} + B_k \times Q_{red}\{k\}\right) \qquad (12.41)$$

where $\Pi P_{sav}\{ls,g,i\}$ is the allocation of loss saving to DG on bus I, and μ_i is the set of branches that lie in between bus i and the reference bus.

To assess the economic efficiency of a network, loss saving obtained by Eq. (12.39) should be equal to the loss saving allocated to each distributed generator, and it is expressed as:

$$P_{sav}\{ls\} \approx \sum_{i=1}^{n} \Pi P_{sav}\{ls,g,i\} \qquad (12.42)$$

From Eq. (12.42), it is seen that the left-hand side is approximately equal to the right-hand side due to neglecting the variations of node voltages in the presence of DGs. The error achieved is around 1% to 2% of total load capacity, or occasionally less, which depends on the locations and size of DGs in the system.

12.3 RESULTS AND DISCUSSION

In order to test the effectiveness of the proposed method, a 28-bus radial distribution system was selected. The system's single-line diagram is shown in Figure 12.3, and each bus, except bus 1, is connected to a load. Practical line and load data have been obtained from [21]. The base values for per-unit calculations are 1 MVA and 11 kV for the given test system. The system has been modified by suboptimally placing

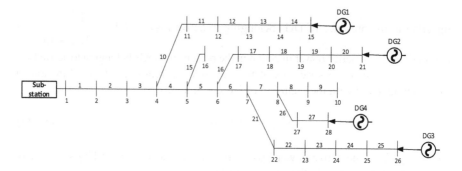

FIGURE 12.3   Single-line diagram of a 28-bus distribution system with DGs

and sizing DGs. DGs have been assumed to be located suboptimally at buses 15, 21, 26, and 28 for electrical connection. DG ratings are typically determined by local consumers' power consumption. Therefore, the active power of DGs has been determined to be 0.70 times the respective consumer's power consumption, and the reactive power of DGs has been determined to be 0.35 times the respective consumer's power consumption. The proposed algorithm takes into account consumer PFs as well as DGs. PF variations within a specified range of 0.65, 0.75, 0.85, 0.95, and 0.985 have been considered for analysis as they are commonly observed in practical distribution systems. The reference PF for the given test system is set to 0.85.

Figure 12.4 shows that the active and reactive power consumption by consumers varies with different PFs. Table 12.1 presents the changes in active and reactive power consumption of consumers when the PF is varied with respect to the reference PF. The results indicate that as the PF increases gradually from 0.65 to 0.985, consumers can draw more active power while consuming less reactive power. This means that at a PF of 0.985, consumers can increase their active power consumption while decreasing their reactive power consumption, thereby reducing the reactive power loading and increasing the active power loading on the distribution lines.

As the PF improves, the change in active power consumption is comparatively smaller than that of reactive power consumption. For instance, at 0.985 PF, there is a positive change of 64.238% in active power consumption compared to that at 0.65 PF, where there is a negative change of 81.384% in reactive power consumption. This indicates that with an improvement in PF from 0.65 to 0.985, the load can draw 64.238% more active power while reducing its reactive power consumption by 81.384%.

Table 12.2 shows the active loss allocation of consumers on various PFs. Active system losses are 103.36 kW, 68.819 kW, and 46.622 kW on 0.65 PF, 0.85 PF, and 0.985 PF, respectively. There is an approximately 55% reduction in system losses when PF switches to 0.985 from 0.65. It does, however, have an immediate effect on loss allocation. If the PF of the consumer is improved, active loss allocation to consumers is decreased and vice versa.

Table 12.3 shows active and reactive power injection by DGs at suboptimal locations. Table 12.4 shows allocation of active loss saving to DGs. Table 12.3 follows the

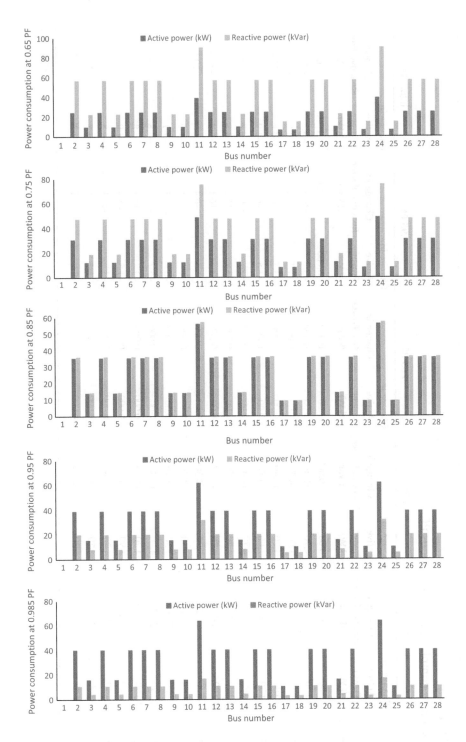

FIGURE 12.4 Complex power consumption by consumers on various PF

TABLE 12.1

Change in complex power on various PF

Bus no.	0.65 PF Change in Active Power (kW)	0.65 PF Change in Reactive Power (kVar)	0.75 PF Change in Active Power (kW)	0.75 PF Change in Reactive Power (kVar)	0.85 PF Change in Active Power (kW)	0.85 PF Change in Reactive Power (kVar)	0.95 PF Change in Active Power (kW)	0.95 PF Change in Reactive Power (kVar)	0.985 PF Change in Active Power (kW)	0.985 PF Change in Reactive Power (kVar)
1	0.00	0.00	0.00	0.000	0.00	0.000	0.000	0.000	0.000	0.000
2	10.90	20.70	4.70	11.40	0.00	0.000	-3.7	-16.1	-4.8	-25.4
3	4.30	8.20	1.90	4.50	0.00	0.000	-1.5	-6.4	-1.9	-10.1
4	10.90	20.70	4.70	11.40	0.00	0.000	-3.7	-16.1	-4.8	-25.4
5	4.30	8.20	1.90	4.50	0.00	0.000	-1.5	-6.4	-1.9	-10.1
6	10.90	20.70	4.70	11.40	0.00	0.000	-3.7	-16.1	-4.8	-25.4
7	10.90	20.70	4.70	11.40	0.00	0.000	-3.7	-16.1	-4.8	-25.4
8	10.90	20.70	4.70	11.40	0.00	0.000	-3.7	-16.1	-4.8	-25.4
9	4.30	8.20	1.90	4.50	0.00	0.000	-1.5	-6.4	-1.9	-10.1
10	4.30	8.20	1.90	4.50	0.00	0.000	-1.5	-6.4	-1.9	-10.1
11	17.20	32.90	7.50	18.10	0.00	0.000	-5.9	-25.5	-7.7	-40.4
12	10.90	20.70	4.70	11.40	0.00	0.000	-3.7	-16.1	-4.8	-25.4
13	10.90	20.70	4.70	11.40	0.000	0.000	-3.7	-16.1	-4.8	-25.4
14	4.30	8.20	1.90	4.50	0.000	0.000	-1.5	-6.4	-1.9	-10.1
15	10.90	20.70	4.70	11.40	0.000	0.000	-3.7	-16.1	-4.8	-25.4
16	10.90	20.70	4.70	11.40	0.000	0.000	-3.7	-16.1	-4.8	-25.4
17	2.80	5.30	1.20	2.90	0.000	0.000	-0.9	-4.1	-1.2	-6.5

(Continued)

TABLE 12.1 (Continued)

Change in complex power on various PF

Bus no.	0.65 PF		0.75 PF		0.85 PF		0.95 PF		0.985 PF	
	Change in Active Power (kW)	Change in Reactive Power (kVar)	Change in Active Power (kW)	Change in Reactive Power (kVar)	Change in Active Power (kW)	Change in Reactive Power (kVar)	Change in Active Power (kW)	Change in Reactive Power (kVar)	Change in Active Power (kW)	Change in Reactive Power (kVar)
18	2.80	5.30	1.20	2.90	0.000	0.000	-0.9	-4.1	-1.2	-6.5
19	10.90	20.70	4.70	11.40	0.000	0.000	-3.7	-16.1	-4.8	-25.4
20	10.90	20.70	4.70	11.40	0.000	0.000	-3.7	-16.1	-4.8	-25.4
21	4.30	8.20	1.90	4.50	0.000	0.000	-1.5	-6.4	-1.9	-10.1
22	10.90	20.70	4.70	11.40	0.000	0.000	-3.7	-16.1	-4.8	-25.4
23	2.80	5.30	1.20	2.90	0.000	0.000	-0.9	-4.1	-1.2	-6.5
24	17.20	32.90	7.50	18.10	0.000	0.000	-5.9	-25.5	-7.7	-40.4
25	2.80	5.30	1.20	2.90	0.000	0.000	-0.9	-4.1	-1.2	-6.5
26	10.90	20.70	4.70	11.40	0.000	0.000	-3.7	-16.1	-4.8	-25.4
27	10.90	20.70	4.70	11.40	0.000	0.000	-3.7	-16.1	-4.8	-25.4
28	10.90	20.70	4.70	11.40	0.000	0.000	-3.7	-16.1	-4.8	-25.4

TABLE 12.2

Active loss allocation (kW) of consumers

Bus no.	0.65 PF			0.85 PF			0.985 PF		
	Proposed Method	Proportional Method	Quadratic Method	Proposed Method	Proportional Method	Quadratic Method	Proposed Method	Proportional Method	Quadratic Method
1	0.000	0.000	0.000	0.000	0.000	0.000	0.000	0.000	0.000
2	0.890	0.900	0.894	0.591	0.599	0.595	0.395	0.406	0.403
3	0.879	0.531	0.321	0.583	0.354	0.214	0.390	0.24	0.145
4	3.201	3.271	3.287	2.119	2.178	2.189	1.417	1.476	1.483
5	1.678	1.032	0.641	1.109	0.687	0.427	0.74	0.466	0.289
6	5.014	5.19	5.281	3.311	3.456	3.516	2.208	2.341	2.382
7	5.717	5.92	6.035	3.772	3.942	4.018	2.513	2.67	2.722
8	5.909	6.129	6.261	3.897	4.081	4.168	2.596	2.765	2.824
9	2.354	1.468	0.928	1.553	0.977	0.618	1.034	0.662	0.419
10	2.362	1.475	0.936	1.558	0.982	0.623	1.037	0.665	0.422
11	5.919	7.273	8.288	3.915	4.843	5.518	2.617	3.281	3.738
12	3.882	3.94	3.95	2.567	2.623	2.630	1.716	1.777	1.782
13	3.974	4.038	4.054	2.627	2.688	2.699	1.756	1.821	1.829
14	1.587	0.955	0.573	1.049	0.636	0.381	0.701	0.431	0.258
15	4.019	4.086	4.105	2.657	2.72	2.733	1.776	1.843	1.851
16	4.327	4.456	4.515	2.86	2.967	3.006	1.909	2.01	2.036
17	1.315	0.602	0.279	0.868	0.401	0.186	0.579	0.272	0.126
18	1.337	0.614	0.285	0.883	0.409	0.19	0.588	0.277	0.129
19	5.418	5.635	5.761	3.575	3.752	3.836	2.383	2.541	2.598
20	5.497	5.722	5.856	3.626	3.81	3.899	2.417	2.581	2.641

(Continued)

TABLE 12.2 (Continued)

Active loss allocation (kW) of consumers

Bus no.	0.65 PF			0.85 PF			0.985 PF		
	Proposed Method	Proportional Method	Quadratic Method	Proposed Method	Proportional Method	Quadratic Method	Proposed Method	Proportional Method	Quadratic Method
21	2.205	1.387	0.89	1.455	0.924	0.593	0.969	0.626	0.402
22	5.984	6.183	6.294	3.945	4.117	4.190	2.628	2.789	2.839
23	1.556	0.708	0.327	1.026	0.471	0.218	0.683	0.319	0.148
24	9.899	12.292	14.067	6.523	8.184	9.366	4.343	5.544	6.345
25	1.590	0.722	0.334	1.048	0.481	0.222	0.698	0.326	0.151
26	6.276	6.468	6.572	4.136	4.306	4.376	2.753	2.917	2.964
27	5.955	6.175	6.308	3.927	4.112	4.200	2.615	2.785	2.845
28	5.967	6.187	6.32	3.935	4.119	4.208	2.620	2.791	2.851

TABLE 12.3

Complex power injections by distributed generators

	0.65 PF		0.75 PF		0.85 PF		0.95 PF		0.985 PF	
Bus no.	Active Power (kW)	Reactive Power (kVar)	Active Power (kW)	Reactive Power (kVar)	Active Power (kW)	Reactive Power (kVar)	Active Power (kW)	Reactive Power (kVar)	Active Power (kW)	Reactive Power (kVar)
15	17.097	19.856	21.403	16.594	24.696	12.597	27.296	6.977	28.081	3.696
21	6.785	7.879	8.493	6.585	9.8	4.999	10.832	2.769	11.143	1.467
26	17.097	19.856	21.403	16.594	24.696	12.597	27.296	6.977	28.081	3.696
28	17.097	19.856	21.403	16.594	24.696	12.597	27.296	6.977	28.081	3.696

TABLE 12.4
Loss savings (kW) attributable to distributed generators

Bus no.	0.65 PF			0.85 PF			0.985 PF		
	Proposed Method	Proportional Method	Quadratic Method	Proposed Method	Proportional Method	Quadratic Method	Proposed Method	Proportional Method	Quadratic Method
15	3.302	3.996	4.102	2.753	3.308	3.389	2.323	2.792	2.848
21	1.775	1.513	1.523	1.492	1.245	1.216	1.273	1.053	0.995
26	4.947	4.662	4.591	4.154	3.945	3.909	3.545	3.374	3.367
28	4.740	4.594	4.548	3.980	3.881	3.866	3.394	3.316	3.325

same trend as Figure 12.4. While incrementally increasing the PF, DGs contribute to the localization of more complex power, and as a result, they receive more revenue for active power injection than from reactive power injection. As illustrated in Table 12.4, the additional economic benefits of DGs are obtained by assigning the loss saving associated with their power injection locally.

Since the PF is changing in increments from 0.65 to 0.985, all DGs are offered greater revenue. It is evident that increasing PF provides an economic advantage for DGs by enabling more active power injection and commensurate loss savings. From the preceding discussion, it can be seen that improving PF delivers technical and financial benefits not only to consumers, but also to DGs seeking to optimize their economic gains.

12.4 CONCLUSION

The chapter presents a new method for allocating system losses and loss savings in deregulated electricity markets, considering consumer PFs and distributed generators. The proposed method uses a branch-oriented approach and applies the electrical network principle for power flow calculations. It also considers the dynamic changes of both active and reactive power in response to PF variations to allocate system losses and savings associated with distributed generator power injections. The chapter also addresses the issue of consumer cross subsidies by dividing participants into two groups and identifying each participant's contribution to system losses within each group. This method can help reduce the economic burden on participants. Moreover, it allows distributed generators to obtain additional economic benefits by injecting both active and reactive power in response to variations in consumer PF. Overall, the chapter's contributions provide a more efficient and fair mechanism for allocating system losses and loss savings in deregulated electricity markets.

REFERENCES

1. Savier, J.S., and Das, D. An exact method for loss allocation in radial distribution systems. *Int J Electr Power Energy Syst* 2012; 36(1): 100–106.
2. Kapitonov, I.A., and Batyrova, N.T. Economic prospects of replacing distributed generation with biofuel. *Int J Energy Research* 2021; 45(12): 17502–17514.
3. Yang, B., Yu, L., Chen, Y., Ye, H., Shao, R., Shu, H., Yu, T., Zhang, X., and Sun, L. Modelling, applications, and evaluations of optimal sizing and placement of distributed generations: A critical state-of-the-art survey. *Int J Energy Research* 2020; 45(3): 3615–3642.
4. Leghari, Z.H., Hassan, M.Y., Said, D.M., Memon, Z.A., and Hussain, Saddam. An efficient framework for integrating distributed generation and capacitor units for simultaneous grid-connected and islanded network operations. *Int J Energy Research* 2021; 45(10): 14920–14958.
5. Nikolaidis, A.I., Charalambous, C.A., and Mancarella, P. A graph-based loss allocation framework for transactive energy markets in unbalanced radial distribution networks. *IEEE Trans Power Syst* 2019; 34(5): 4109–4118.
6. Usman, M., Coppo, M., Bignucolo, F., Turri, R., and Cerretti, A. Multi-phase losses allocation method for active distribution networks based on branch current decomposition. *Int J Electr Power Energy Syst* 2019; 110: 613–622.

7. Moret, F., Tosatto, A., Baroche, T., Pinson, P. Loss allocation in joint transmission and distribution peer-to-peer markets. *IEEE Trans Power Syst* 2021; 36(3): 1833–1842.

8. Kumar, Pankaj, Gupta, N., Niazi, K.R., and Swarnkar, A. Branch current decomposition method for loss allocation in contemporary distribution systems. *Int J Electr Power Energy Syst* 2018; 99: 134–145.

9. Shafeeque Ahmed, K., and Karthikeyan, S.P. Modified penalized quoted cost method for transmission loss allocation including reactive power demand in deregulated electricity market. *Sust Energy, Grids Networks* 2018; 16: 370–379.

10. Zarabadipoura, H., and Mahmoudib, H. A novel loss allocation in pool markets using weight-based sharing and voltage sensitivity analysis. *Electr Power Syst Research* 2017; 152: 84–91.

11. Moon, Y.H., Yun, K.H., Joo, W., and Yang, B.M. Advanced path integral algorithm of transmission loss allocation using smoothness of loss sensitivity. *IFAC-PapersOnLine* 2015; 48(30): 257–263.

12. Khosravi, M., Monsef, H., and Aliabadi, M.H. Loss allocation in distribution network including distributed energy resources (DERs). *Int Trans Electr Energy Syst* 2018; 28(6): e2548.

13. Alayande, A.S., Jimoh, A.A., Yusuff, A.A. An alternative algorithm for solving generation-to-load matching and loss allocation problems. *Int Trans Electr Energy Syst* 2017; 27(8): e2347.

14. Hota, A.P., and Mishra, S. A forward-backward sweep based numerical approach for active power loss allocation of radial distribution network with distributed generations. *Int J Numer Model: Electr Networks, Devices Fields* 2020; 34(1): 1–29.

15. Yu, Q., Xie, J., Chen, X., Yu, K., Gan, L., and Chen, L. Loss allocation for radial distribution networks including DGs using Shapley value sampling estimation. *IET Gener Trans Distrib* 2019; 13(8): 1382–1390.

16. Yu, Q., Xie, J., Chen, X., Yu, K., and Gan, L. Loss and emission reduction allocation in distribution networks using MCRS method and Aumann–Shapley value method. *IET Gener Trans Distrib* 2018; 12(22): 5975–5981.

17. Amaris, H., Molina, Y.P., Alonso, M., and Luyo, J.E. Loss allocation in distribution networks based on Aumann–Shapley. *IEEE Trans Power Syst* 2018; 33(6): 6655–6666.

18. Al-Digs, A., Chen, Y.C. Power system loss divider. *IEEE Trans Power Syst* 2020; 35(4): 3286–3289.

19. Savier, J.S., Das, D. Loss allocation to consumers before and after reconfiguration of radial distribution networks. *Int J Electr Power Energy Syst* 2011; 33: 540–549.

20. Ghofrani-Jahromi, Z., Mahmoodzadeh, Z., and Ehsan, M. Distribution loss allocation for radial systems including DGs. *IEEE Trans Power Syst* 2014; 29(1): 72–80.

21. Das, D., Nagi, H.S., and Kothari, D.P. Novel method for solving radial distribution networks. *IEE Proc Gener Transm Distrib* 1994; 141(4): 291–298.

7. Bialek J., Ziemianek S., and Abi-Samra N., Cross-border transit flow calculation in real interconnected and distribution power networks. IARR Tech. Power Sys. 20 (2): 1817–1824, 2012.

8. Sharma, Pankaj, Gupta, N., Niazi, K. R., and Swarnkar, A. Branch current decomposition method for loss allocation in contemporary distribution systems. Int. J. Electr. Power.
Energy Syst. 99: 134–145, 2018.

9. Atanasovski, Metodija, and Kerngo-Tenekova R. M. Method for loss allocation for radial distribution systems, including reactive power, submitted to some distributed energy. IEEE Trans. Power Apparatus. 2016.

10. Kumar, V. K., Hn, and Khanumaje, H.V. Novel loss allocation with peak shaving and peak valleam sharing and voltage profile on by analyzing. Int. J. Electr. Power Energy Syst. 2017.

11. Savier, V.H., Yan, K.H., Jan, W., and Kang, K.M. A scheme for separating the transmission loss allocation using consequence of sensitivity. IEEE Power Del. 2015. 2015(3) 241–251.

12. Kaspirek, M., Mezera, D., and Alhaddad, M.H. Loss allocation in a distribution system including distributed energy resources (DER). IEEE Trans. Energy Syst. 2015. 2015.

13. Shi, Belch, A.S., etc., Newman A.L. A consideration for solution to some long centralized machine and classification problems. Adv. Electr. & Co. Energy Syst. 2017. 2015. 2015(3).

14. Hota, A.P., and Naik, S.A. Issue HPC and some novel computation of model for sizing power location for electrical distribution system with distributed energy resources. Int. J. Numer. Model. Electr. Networks, Devices Fields. 2015(3).

15. Sheng, Min, J., Chen, X., Yu, K., Tan, L., and Chen. C. Loss allocation in a distribution network including loss-making flexible value, using the economic classifier. Energy Build. 2015. 2015(3)–128.

16. Wu, D., Bao, J., Sun, X., Yang, L., and Chen, L. Loss and value value breakdown allocation in a consumer power network for RES method and Amperes Sharing. Alternating & RET power Technol. 2015(3), 2015–2031.

17. Alonso, D., Seltner, V.H., Alonso, M., and Leon, J.P. Loss Aranda analysis consumers under electrical system with lag. IEEE Trans. Power Sys. New computation electric under power. Comp.. Chem. Eng. Power system net divide., IEEE Trans. Power Sys. 2015. 200–280.

18. Savier, J.S., Das, D. Loss allocation to consumers before and after reorganization of radial distribution networks. Int. J. Electr. Power Energy Syst. 2012. 2012(2), 540–549.

19. Chartrand-Pelton, J., and Chamberline, Z. Applied Mathematical optimization for radial systems including DGs. IEEE Trans. Power Sys. 2015. 2015.

20. Das, D., Nagi, H.S., and Kothari, D.P. Novel method for solving radial distribution networks. IEE Proc. Gener. Transm. Distrib. 1994. 141(4): 291–298.

Index

For Product Safety Concerns and Information please contact our
EU representative GPSR@taylorandfrancis.com, Taylor & Francis
Verlag GmbH, Kaufingerstraße 24, 80331 München, Germany